세상에서 가장 쉬운 과학 수업

별의 물리학

세상에서 가장 쉬운 과학 수업
별의 물리학
ⓒ 정완상, 2024

초판 1쇄 인쇄 2024년 7월 1일
초판 1쇄 발행 2024년 7월 17일

지은이 정완상
펴낸이 이성림
펴낸곳 성림북스

책임편집 최윤정
디자인 쏘울기획

출판등록 2014년 9월 3일 제25100-2014-000054호
주소 서울시 은평구 연서로3길 12-8, 502
대표전화 02-356-5762
팩스 02-356-5769
이메일 sunglimonebooks@naver.com

ISBN 979-11-93357-30-9 03400

노벨상 수상자들의 **오리지널 논문**으로 배우는 과학

세상에서 가장 쉬운 과학 수업

별의 물리학

정완상 지음

고대 천문학부터 찬드라세카르의 별의 죽음 이론까지
별의 탄생과 진화 그리고 죽음, 그 비밀을 밝히다

성림원북스

CONTENTS

과학을 처음 공부할 때 이런 책이 있었다면 얼마나 좋았을까

남순건(경희대학교 이과대학 물리학과 교수 및 전 부총장)

21세기를 20여 년 지낸 이 시점에서 세상은 또 엄청난 변화를 맞이하리라는 생각이 듭니다. 100년 전 찾아왔던 양자역학은 반도체, 레이저 등을 위시하여 나노의 세계를 인간이 이해하도록 하였고, 120년 전 아인슈타인에 의해 밝혀진 시간과 공간의 원리인 상대성이론은 이 광대한 우주가 어떤 모습으로 만들어져 왔고 앞으로 어떻게 진화할 것인가를 알게 해주었습니다. 게다가 우리가 사용하는 모든 에너지의 근원인 태양에너지를 핵융합을 통해 지구상에서 구현하려는 노력도 상대론에서 나오는 그 유명한 질량–에너지 공식이 있기에 조만간 성과가 있을 것이라 기대하게 되었습니다.

앞으로 올 22세기에는 어떤 세상이 될지 매우 궁금합니다. 특히 인공지능의 한계가 과연 무엇일지, 또한 생로병사와 관련된 생명의 신비가 밝혀져 인간 사회를 어떻게 바꿀지, 우주에서는 어떤 신비로움이 기다리고 있는지, 우리는 불확실성이 가득한 미래를 향해 달려가고 있습니다. 이러한 불확실한 미래를 들여다보는 유리구슬의 역할을 하는 것이 바로 과학적 원리들입니다.

지난 백여 년간의 과학에서의 엄청난 발전들은 세상의 원리를 꿰뚫어보았던 과학자들의 통찰을 통해 우리에게 알려졌습니다. 이런 과학 발전을 가능하게 한 영웅들의 생생한 숨결을 직접 느끼려면 그들이 썼던 논문들을 경험해보는 것이 좋습니다. 그런데 어느 순간 일반인과 과학을 배우는 학생들은 물론, 그 분야에서 연구를 하는 과학자들마저 이런 숨결을 직접 경험하지 못하고 이를 소화해서 정리해 놓은 교과서나 서적들을 통해서만 접하고 있습니다. 창의적인 생각의 흐름을 직접 접하는 것은 그런 생각을 했던 과학자들의 어깨 위에서 더 멀리 바라보고 새로운 발견을 하고자 하는 사람들에게 매우 중요합니다.

저자인 정완상 교수가 새로운 시도로써 이러한 숨결을 우리에게 전해주려 한다고 하여 그의 30년 지기인 저는 매우 기뻤습니다. 그는 대학원생 때부터 당시 혁명기를 지나면서 폭발적인 발전을 하고 있던 끈 이론을 위시한 이론물리학 분야에서 가장 많은 논문을 썼던 사람입니다. 그리고 그러한 에너지가 일반인들과 과학도들을 위한 그의 수많은 서적을 통해 이미 잘 알려져 있습니다. 저자는 이번에 아주 새로운 시도를 하고 있고 이는 어쩌면 우리에게 꼭 필요했던 것일 수 있습니다. 대화체로 과학의 역사와 배경을 매우 재미있게 설명하고, 그 배경 뒤에 나왔던 과학 영웅들의 오리지널 논문들을 풀어간 것입니다. 과학사를 들려주는 책들은 많이 있으나 이처럼 일반인과 과학도의 입장에서 질문하고 이해하는 생각의 흐름을 따라 설명한 책은 없습니다. 게다가 이런 준비를 마친 후에 아인슈타인 같은 영웅들의 논문을 원래의 방식과 표기를 통해 설명하는 부분은 오랫동안 과학을 연구해온 과학자에게도 도움을 줍니다.

이 책을 읽는 독자들은 복 받은 분들일 것이 분명합니다. 제가 과학을 처음 공부할 때 이런 책이 있었다면 얼마나 좋았을까 하는 생각이 듭니다. 정완상 교수는 이제 새로운 형태의 시리즈를 시작하고 있습니다. 독보적인 필력과 독자에게 다가가는 그의 친밀성이 이 시리즈를 통해 재미있고 유익한 과학으로 전해지길 바랍니다. 그리하여 과학을 멀리하는 21세기의 한국인들에게 과학에 대한 붐이 일기를 기대합니다. 22세기를 준비해야 하는 우리에게는 이런 붐이 꼭 있어야 하기 때문입니다.

우리나라 기초과학의 미래가 될 학생들을 위한 책

김현주(한민고등학교 물리 교사)

물리학 수업 첫 번째 시간에 항상 학생들에게 이런 질문을 합니다.

"여러분, '물리학' 하면 무엇이 생각나나요?"

이 질문에 대한 다양한 답변이 나옵니다.

"어려운 과목이요!"

"천재들이 하는 거요."

"만물의 이치요."

대화를 통해 학생들이 알고 있는 배경지식과 물리에 대한 호기심의 깊이, 그리고 기대감을 확인할 수 있습니다.

현재 고등학교 선택 교육과정에서는 우주의 탄생, 원소의 생성, 별의 진화, 특수상대성이론 등 다양한 현대 물리학과 천문학 내용을 배웁니다. 선택 교육과정이다 보니 정말 과학 분야에 관심 있고 배우고자 하는 열정을 가진 학생들이 모여 수업을 듣습니다.

우주론, 원소의 기원, 별의 진화, 핵융합, 핵분열, 상대성이론, 양자역학 등 이론 수업을 할 때 이와 관련된 과학사와 과학자들의 뒷이야기를 하곤 합니다. 그러면 학생들이 재미있어하고 높은 곳에 있는 것처럼 느껴지는 과학자들을 친근하게 받아들입니다.

그러나 물리학과 천문학은 수학과 과학에 대한 기초지식이 있어야 제대로 이해하고 학습할 수 있습니다. 수식을 배제하고 정성적인 내용만을 다루는 것은 학습 동기 유발에는 좋지만, 그 과정인 수식도 이해해야 창의적인 사고를 하고 더 앞으로 나아갈 수 있습니다. 그런 의미에서 수식을 피하지 않고 과학 이론을 설명하는《세상에서 가장 쉬운 과학 수업 별의 물리학》의 기획은 꼭 필요한 것이라고 봅니다.

과학을 좋아하는 학생들과 함께 과학 동아리, STEAM R&E, 과학 과제연구 등 다양한 과학 탐구 활동을 할 때, 학생들이 책을 추천해 달라고 하는 경우가 많습니다. 관심 있고 탐구하고 싶은 분야는 많지만, 아직은 고등학교 과학을 배우는 과정이라서 더 많은 학습 후에 이해 가능한 주제도 있습니다. 학생들이 추가로 찾아보는 자료들은 주로 인터넷 동영상과 블로그입니다. 옳은 내용도 있지만 대개 서로 붙여 넣어 반복되는 내용들이 많아, 심화된 지식을 알고 싶어도 학생들 스스로 자료를 찾기는 쉽지 않습니다. 그런 경우 논문 검색 사이트에서 관련 분야 논문을 찾는 방법을 알려 주고 전공 서적을 추천하지만, 학생들이 이해하기에는 너무 어려운 내용이 대부분입니다.

정완상 교수님의 〈노벨상 수상자들의 오리지널 논문으로 배우는 과학〉 시리즈는 이런 학생들에게 적극 추천하고 싶은 책입니다. 수식과 증명 과정을 학생들이 이해하기 쉽게 기술하였고, 특히 고교생 중 물리학에 관심이 많은 학생이 읽으면 관련 배경지식뿐만 아니라 과학사에서 유명한(이야기로만 전해 들은) 논문들의 원문을 직접 접할 수 있습니다. 또한 그 과정에서 과학자들의 연구 열정과 노력도 알게 됩니다. 이 책이 우리나라 기초과학의 미래가 될 학생들로 하여금 진로 선택을 하는 데 있어서 매우 긍정적인 역할을 할 것이라고 확신합니다.

천재 과학자들의 오리지널 논문을
이해하게 되길 바라며

사람들은 과학 특히 물리학 하면 너무 어렵다고 생각하지요. 제가 외국인들을 만나서 얘기할 때마다 신선하게 느끼는 점이 있습니다. 그들은 고등학교까지 과학을 너무 재미있게 배웠다고 하더군요. 그래서인지 과학에 대해 상당한 지식을 가진 사람들이 많았습니다. 그 덕분에 노벨 과학상도 많이 나오는 게 아닐까 생각해요. 우리나라는 노벨 과학상 수상자가 한 명도 없습니다. 이제 청소년과 일반 독자의 과학 수준을 높여 노벨 과학상 수상자가 매년 나오는 나라가 되게 하고 싶다는 게 제 소망입니다.

그동안 양자역학과 상대성이론에 관한 책은 전 세계적으로 헤아릴 수 없을 정도로 많이 나왔고 앞으로도 계속 나오겠지요. 대부분의 책은 수식을 피하고 관련된 역사 이야기들 중심으로 쓰여 있어요. 제가 보기에는 독자를 고려하여 수식을 너무 배제하는 것 같았습니다. 이제는 독자들의 수준도 많이 높아졌으니 수식을 피하지 말고 천재 과학자들의 오리지널 논문을 이해하길 바랐습니다. 그래서 앞으로 도래할 양자(量子, quantum)와 상대성 우주의 시대를 멋지게 맞이하도록 도우리라는 생각에서 이 기획을 하게 된 것입니다.

원고를 쓰기 위해 논문을 읽고 또 읽으면서 어떻게 이 어려운 논문을 독자들에게 알기 쉽게 설명할까 고민했습니다. 여기서 제가 설정한 독자는 고등학교 정도의 수식을 이해하는 청소년과 일반 독자입니다. 물론 이 시리즈의 논문에 그 수준을 넘어서는 내용도 나오지만 고등학교 수학만 알면 이해할 수 있도록 설명했습니다. 이 책을 읽으며 천재 과학자들의 오리지널 논문을 얼마나 이해할지는 독자들에 따라 다를 거라 생각합니다. 책을 다 읽고 100% 혹은 70%를 이해하거나 30% 미만으로 이해하는 독자도 있을 것입니다. 저의 생각으로는 이 책의 30% 이상 이해한다면 그 사람은 대단하다고 봅니다.

이 책에서 저는 별의 물리에 관한 몇 편의 논문을 다루었습니다. 별 이론에 대한 에딩턴과 레인의 논문 및 별의 탄생 과정을 논한 베테의 논문, 별의 죽음을 연구한 찬드라세카르의 논문이 그것입니다.

20세기 위대한 업적인 별에 관한 연구를 독자들이 쉽게 이해할 수 있도록 고대 사람들이 생각했던 별 이야기를 곁들였습니다. 뉴턴의 만유인력, 별빛에 대한 광행차 현상과 별까지의 거리를 측정하는 방법, 별의 절대등급에 대한 역사도 살펴보았습니다.

또한 별이 어떻게 안정된 형태를 유지하는지 알려주는 레인-엠덴 방정식을 자세하게 설명했습니다. 이를 위해 간단하게 열역학을 다루면서 열역학 과정 중 하나인 폴리트로픽 과정과 이를 이용한 레인과 엠덴의 폴리트로픽 별 모형을 소개했습니다.

별의 등급과 분류에 대한 연구의 역사를 가지고 별의 일생도 알아보았습니다. 마지막으로 별의 죽음의 한 형태인 백색 왜성의 물리학을 논한 찬드라세카르의 논문을 다루었습니다. 이 책을 통해 여러분이 레인-엠덴 방정식이나 아주 짧은 논문인 찬드라세카르의 논문을 이해하리라 여깁니다.

〈노벨상 수상자들의 오리지널 논문으로 배우는 과학〉 시리즈는 많은 이에게 도움을 줄 수 있다고 생각합니다. 과학자가 꿈인 학생과 그의 부모, 어릴 때부터 수학과 과학을 사랑했던 어른, 양자역학과 상대성이론을 좀 더 알고 싶은 사람, 아이들에게 위대한 논문을 소개하려는 과학 선생님, 반도체나 양자 암호 시스템, 우주 항공 계통 등의 일에 종사하는 직장인, 〈인터스텔라〉를 능가하는 SF 영화를 만들고 싶어 하는 영화 제작자나 웹툰 작가 등 많은 사람들에게 이 시리즈를 추천합니다.

진주에서 정완상 교수

세상에서 가장 쉬운 과학 수업 별의 물리학

별의 죽음 과정을 물리학적으로 밝히다
_ 겐첼 박사 깜짝 인터뷰

별이 죽는다는 것

기자   오늘은 별의 죽음 과정을 물리학적으로 밝힌 찬드라세카르의 논문에 대해 겐첼 박사와 인터뷰를 진행하겠습니다. 겐첼 박사는 2020년 우리은하 중심에서 초대형 블랙홀을 발견해 노벨 물리학상을 수상한 분이지요. 겐첼 박사님, 나와 주셔서 감사합니다.

겐첼   제가 제일 존경하는 과학자인 찬드라세카르의 논문에 관한 내용이라 만사를 제치고 달려왔습니다.

기자   별이 죽는 것은 어떤 의미죠?

겐첼   별은 성간물질이 뭉쳐서 만들어집니다. 이를 원시별이라고 부르지요. 원시별 속에서는 핵융합이 일어나는데 이것이 바로 별이 열과 빛을 내는 이유입니다. 원시별은 점점 커졌다가 더 이상 핵융합이 일어나지 않으면 죽게 되지요. 캠프파이어의 장작을 생각해 보세요. 장작을 태우면 열과 빛이 나옵니다. 하지만 장작이 다 타 버리면 더는 열과 빛이 나오지 않습니다.

기자   이해가 되네요.

찬드라세카르 이전의 별의 물리

기자   찬드라세카르 이전에 이루어진 별의 연구에 대해 알려 주세요.

겐첼   별의 구조 방정식을 알아낸 레인과 엠덴의 연구가 최초로 별을 물리학적으로 다룬 것이지요. 그들은 별 속 기체의 압력과 중력이 평형이 되는 조건을 통해 별의 평형 방정식(레인-엠덴 방정식)을 발견했어요. 레인-엠덴 방정식은 별 내부의 기체 압력과 중력이 평형을 이루는 모형인데, 영국 물리학자 에딩턴은 이 방정식에 별 내부의 복사 압력을 추가해야 한다고 생각했지요. 그 후 캐넌이 별을 표면 온도에 따라 O, B, A, F, G, K, M으로 분류했고, 덴마크의 헤르츠스프룽과 러셀이 별의 밝기와 온도의 관계를 조사했습니다. 그리고 진스가 원시별 이론을, 베테가 별 탄생 이론을 연구했지요.

기자   많은 과학자가 별의 물리학을 연구하고 있었군요.

겐첼   그렇습니다.

찬드라세카르 논문의 개요

기자   찬드라세카르의 논문에는 무슨 내용이 담겨 있나요?

겐첼   별이 죽는 모습은 별의 질량에 따라 다릅니다. 태양처럼 가벼운 별은 백색 왜성으로 변하고 그보다 무거운 별은 중성자별로 변합니다. 더 무거워지면 블랙홀이 만들어지지요. 찬드라세카르는 태양처럼 가벼운 별이 죽어서 만들어지는 백색 왜성에 대한 평형 조건을 발견했습니다. 즉, 백색 왜성의 질량과 반지름 사이의 관계를 찾아낸 거지요.

기자　어떻게 발견한 거죠?

겐첼　찬드라세카르는 백색 왜성의 전체 압력이 복사 압력과 기체 압력의 합이라고 생각했어요. 그리고 백색 왜성은 죽은 별이므로 별의 복사 에너지가 너무 작아 무시할 수 있다고 보았어요. 비록 죽은 별이지만 백색 왜성 속의 전자는 축퇴 에너지를 가지며, 그 에너지가 압력을 만들어 중력에 대한 압력과 평형을 유지해 백색 왜성을 공 모양으로 유지한다고 생각했지요. 이 때문에 찬드라세카르는 백색 왜성 속의 전자를 양자역학적으로 다루어야 한다고 판단했습니다. 그는 별의 중심부의 중력에 의한 압력과 전자의 축퇴 압력이 같다는 조건으로부터 백색 왜성의 평형 관계식을 찾아냈지요.

기자　논문을 자세히 살펴보고 싶군요.

찬드라세카르의 논문이 일으킨 파장

기자　찬드라세카르의 논문은 어떤 파장을 몰고 왔나요?

겐첼　베테의 별 탄생 이론부터 찬드라세카르의 별의 죽음 이론까지 완성되면서 별의 물리학이 확립되었어요. 이 이론은 수정을 거쳐서 우리가 백색 왜성이나 중성자별을 관측하고 분류하는 데 크게 기여했지요.

기자   엄청나게 중요한 역할을 했군요. 찬드라세카르 이후에는 별에 대해 무슨 연구들이 진행되었죠?

겐첼   찬드라세카르의 논문은 고전역학에 대한 양자역학 버전을 이용한 계산입니다. 그 후 상대성이론을 적용한 상대론적 양자역학을 가지고 그의 논문이 수정되었습니다. 중성자별의 물리학 연구도 활발히 이루어졌습니다. 많은 과학자가 별의 죽음의 기묘한 형태인 블랙홀을 연구하고, 퀘이사에 관해서도 물리학을 만들려고 노력하고 있습니다. 이렇게 우주를 구성하는 여러 종류의 별에 대한 물리학으로 우리는 우주를 좀 더 이해하게 되었지요. 그 시작이 바로 찬드라세카르의 논문이었습니다.

기자   그렇군요. 지금까지 찬드라세카르의 별의 죽음 논문에 대해 겐첼 박사의 이야기를 들어 보았습니다.

세상에서 가장 쉬운 과학 수업 별의 물리학

첫 번째 만남

•

고대의 별 이론

고대 이집트의 태양력_고정된 별과 방황하는 별

정교수 이 책에서는 별의 물리학을 다룰 거야. 즉, 별의 탄생과 진화, 죽음에 대한 이야기를 할 거라네. 첫 번째로 옛날 사람들은 별을 어떻게 생각했는지 알아보세. 별과 태양, 지구, 달을 최초로 연구한 건 고대 이집트 사람들이지.

물리군 고대라면 언제쯤이죠?

정교수 지금으로부터 4천 년도 더 거슬러 올라가야 해.

물리군 굉장히 오래전이군요.

정교수 별의 정의부터 얘기할까? 별은 스스로 빛을 내는 천체를 말해. 이것을 항성이라고도 부르지. 행성과 위성, 소행성 등은 스스로 빛을 내지 못하는 천체이므로 항성이 아니라네.

물리군 행성은 별 주위를 도는 천체이고, 위성은 행성의 주위를 도는 천체이지요?

정교수 맞아.

물리군 아주 오랜 옛날에는 별과 행성을 어떻게 구별했나요?

정교수 고대 천문학자들은 별과 행성을 명확히 구별할 수 없었어. 대신에 그들은 별을 두 종류로 나누었네. 위치가 변하지 않는 '고정된 별'과 며칠 또는 몇 주에 걸쳐 고정된 별에 비해 눈에 띄게 움직이는 '방황하는 별'이 존재한다고 생각했지.

물리군 '고정된 별'이 별(항성)이고 '방황하는 별'이 행성이군요!

정교수 바로 그걸세. 역사적으로 별은 전 세계 문명에 중요한 영향을 끼쳤어. 항법과 방향을 정하고, 계절의 흐름을 표시하고, 달력을 정의하는 데 사용되었지. 고대 천문학자들은 또한 태양과 별을 구별했다네. 물론 우리는 태양도 별이라는 것을 알지만 말이야.

고대 천문학자들은 별이 지구를 중심으로 하는 거대한 공 모양의 안쪽 표면에 붙어 있다고 생각했다. 이 거대한 공을 천구라고 부른다. 그들은 별의 위치가 달라지는 이유에 대해 천구가 지구를 중심으로 돌기 때문에 천구에 붙어 있는 별들이 돈다고 여겼다.

고대 이집트의 제10왕조(B.C.2130~B.C.2040) 때에는 지평선에서 연속적으로 떠오르는 36개의 항성을 데칸(decan)이라고 불렀는데, 이것을 항성 시계로 사용했다. 열흘마다 새로운 데칸이 태양과 함께 나타났기 때문에 이집트인들에게 데칸은 태양력의 분할을 표시하는

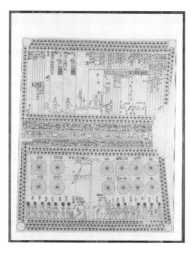

데칸에 관한 이집트인들의 기록

데 쓰였다. 이집트인들은 데칸을 이용해 계절에 따라 낮과 밤의 길이가 달라지는 것도 설명했다.

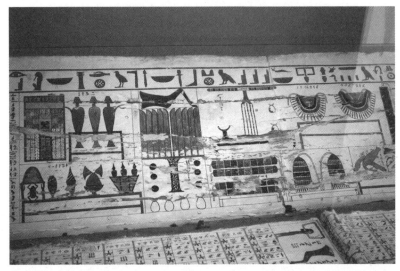

고대 이집트 제11왕조(B.C.2150~B.C.1991)에 작성된 별의 목록(출처: Einsamer Schütze/Wikimedia Commons)

기원전 18세기경 이집트인들은 태양력을 만들었다. 그들은 1년을 365일로 정하고, 한 달을 30일로 하는 12달과 연말에 5일을 더하는 식으로 달력을 완성했다. 또한 나일강이 범람할 때면 동쪽 하늘의 일정한 위치에 시리우스가 떠오르는 것을 알아냈다.

세상에서 가장 쉬운 과학 수업 별의 물리학

별자리_사물, 동물, 신화 속 영웅의 이름을 붙이다

물리군   별자리는 누가 처음 생각했나요?

정교수   별자리(constellation)는 천구의 밝은 별을 중심으로 지구에서 보이는 모습에 따라 이어서 어떤 사물을 연상하도록 이름 붙인 것을 말해. 우리가 사용하는 별자리 이름은 거의 대부분 고대 그리스 시대에 붙여졌어.

대략 기원전 3000년경 바빌로니아 부근에서 별자리를 만들기 시작한 것으로 여겨진다. 별자리 중에서 최초로 발견된 것은 황소자리다.

황소자리

황소자리에서 가장 밝은 별은 1등성 알데바란이다. 알데바란 근처의 별들을 'V'자 모양으로 잇고, 각 끄트머리를 길게 연장해서 하늘에 'Y'자 모양을 만들면 황소자리가 된다.

고대 바빌로니아에서는 황소자리를 '서쪽 하늘 별들의 지도자'로 불렀다. 황소자리는 기원전 12000~15000년경의 구석기 시대 유적지인 라스코 동굴벽화나 괴베클리 테페 등지에서도 발견될 정도로 오랜 역사를 갖고 있다.

바빌로니아 천문학자들의 별자리는 그리스 로마에게로 전승되었다. 그리스 로마 사람들은 별자리에 자신들의 신화에 나온 영웅, 동물 등의 이름을 붙였다.

기원전 8세기경 호메로스와 헤시오도스의 작품에 오리온자리와 큰곰자리 등이 나온다. 기원전 3세기경 시인 아라토스는 〈파이노메나〉에서 44개의 별자리 목록을 소개했다. 이후 대략 사오십 개로 별자리 수가 고정되어 갔다. 현재 쓰이는 별자리는 2세기 후반 그리스 천문학자 프톨레마이오스가 정리한 48개를 기원으로 한다.

예를 들어 물병자리는 그리스 신화에서 물병을 든 가니메데의 모습을 하고 있다. 제우스는 아름다운 소년 가니메데를 보고 사랑에 빠

물병자리

세상에서 가장 쉬운 과학 수업 별의 물리학

져, 그를 올림포스산으로 데리고 가 신들에게 포도주 따르는 일을 시켰다고 한다.

오리온자리는 그리스 신화에 나오는 사냥꾼인 오리온에 기원한다.

오리온자리

독수리자리의 명칭인 '아퀼라(Aquila)'는 제우스에 속한 새의 이름을 딴 것이다. 아퀼라는 제우스가 던진 번개를 되찾아오는 데 이용되었다.

독수리자리

그 외에도 동물을 본뜬 자리로는 사자자리, 큰곰자리 등이 있다.

사자자리

큰곰자리

황도 12궁과 점성술_태양이 지나가는 12개의 별자리

정교수 이번에는 황도 12궁과 점성술에 대해 이야기할게.

물리군 점성술이라면 별자리로 점을 치는 걸 말하나요?

세상에서 가장 쉬운 과학 수업 별의 물리학

정교수　맞아. 황도는 태양이 1년에 걸쳐 하늘을 이동하는 경로를 뜻
해. 고대 천문학자들은 태양이 지구 주위를 돈다고 생각했으니까 말일
세. 이때 태양이 지나가는 12개의 별자리를 황도 12궁이라고 부르지.

기원전 1세기의 황도 12궁
(이집트 덴데라 신전)

6세기의 황도 12궁

태양이 지나가는 12개의 별자리를 날짜에 따라 써 보면 다음과 같다.

양자리	4월 19일 ~ 5월 13일	천칭자리	10월 31일 ~ 11월 22일
황소자리	5월 14일 ~ 6월 19일	전갈자리	11월 23일 ~ 11월 29일
쌍둥이자리	6월 20일 ~ 7월 20일	사수자리	12월 18일 ~ 1월 18일
게자리	7월 21일 ~ 8월 9일	염소자리	1월 19일 ~ 2월 15일
사자자리	8월 10일 ~ 9월 15일	물병자리	2월 16일 ~ 3월 11일
처녀자리	9월 16일 ~ 10월 30일	물고기자리	3월 12일 ~ 4월 18일

황도의 12개 별자리

　　　　　세상에서 가장 쉬운 과학 수업 별의 물리학

황도 12궁과 태양과 별의 운동을 관측해서 사람이나 국가의 미래를 예측하는 것을 점성술이라고 하는데 과학적인 근거는 없다.

태양신 라 이야기_신성한 존재로 여겨진 태양

물리군 고대 이집트 사람들은 태양과 별을 서로 다른 천체로 생각했겠네요.

정교수 그래. 고대 이집트 신화를 보면 그들이 태양을 얼마나 신성한 존재로 여겼는지 알 수 있어. 한번 자세히 살펴볼까?

창조의 신 아톤　　　태양신 라　　　마트　　　세크메트　　　투트
(출처: Jeff Dahl/Wikimedia Commons)

태초에 혼돈의 바다 누에서 벤벤이라는 언덕이 솟아올라, 최초의 신인 아톤이 스스로 태어났다. 아톤은 빛을 창조하였는데, 이 빛은 태양으로서 라가 되었다.

라는 스스로 4명의 딸을 낳았다. 그중 하나인 마트는 태양의 돛단배에 거주하며, 달의 신 투트의 아내로서 진리와 지혜, 정의의 여신으로 추앙받았다.

라의 또 다른 딸은 암사자 머리를 한 파괴의 여신 세크메트이다. 여성적인 힘의 화신이자 전쟁과 복수의 여신 세크메트는 라가 세상을 파괴하기 위해 만들었다. 세크메트는 인류에게 질병과 재앙을 가져다주는 공포의 여신이었다.

마트의 남편인 투트는 달의 신이다. 주로 아프리카검은따오기나 비비의 머리에 사람의 몸을 한 모습으로 묘사된다.

태양신 라는 태양의 돛단배(혹은 태양의 배)를 타고 천상의 나일강, 즉 은하수를 따라 항해한다. 라가 태양의 돛단배를 타고 하늘의 여신인 누트를 따라 동에서 서로 향하면 낮이 되고, 서쪽에서 땅의 신 게브 아래로 내려가 다음 날 동쪽 땅에서 나타나기까지는 밤이 된다.

태양의 돛단배

고대 이집트 사람들은 태양이 지는 것을 죽음, 태양이 뜨는 것은 부활로 보아 서쪽은 죽음의 땅, 동쪽은 부활의 땅으로 생각했다. 그래

세상에서 가장 쉬운 과학 수업 별의 물리학

서 피라미드를 비롯한 대부분의 무덤은 나일강 서편에 짓는 것이 원칙이었다. 부활에 대한 믿음으로 후에 부활할 시체가 훼손되지 않게 방부하는 기술(미라)도 발전했다.

고대 이집트 미라의 관

서쪽 지평선 끝의 은하수에는 마누라는 지역이 있는데, 여기에 도달한 태양의 돛단배는 육체와 분리된 수많은 혼령을 싣고 두아트라는 계곡을 지난다. 이곳에는 거대한 독사 아펩이 있어서 태양신 라는 늘 이 뱀과 싸워야 했으며, 뱀과의 싸움에 져서 잡아먹히면 일식이 찾아온다고 믿었다. 하지만 태양은 부활하기 때문에 아침이 되면 해가 다시 떠오르며 이 싸움은 끝이 없는 것으로 간주되었다.

아펩(출처: Eternal Space/Wikimedia Commons)

물리군 황당하지만 재미있는 신화군요.

정교수 고대 이집트 사람들이 태양과 달을 얼마나 신성하게 여겼는
지를 보여주는 이야기지.

물리군 그렇네요.

탈레스와 아낙시만드로스의 별에 대한 생각_태양, 달, 별은 무엇으로 이루어져 있을까?

정교수 이제 고대 그리스 철학자들의 별과 지구와 달에 대한 생각을
들려주겠네. 먼저 별을 최초로 언급했던 그리스 철학자 탈레스를 알
아보세.

탈레스(Thales of Miletus, B.C.624?~B.C.548?)

세상에서 가장 쉬운 과학 수업 별의 물리학

고대 천문학자들은 지구가 평평하고 그 가장자리로 가면 낭떠러지가 있을 것으로 믿었다. 즉, 해가 서쪽으로 넘어가면 보이지 않는 이유는 낭떠러지 아래로 떨어지기 때문으로 추측했다.

탈레스는 지중해를 항해하면서 관찰한 결과를 가지고 지구가 거대한 바다 위에 떠 있는 납작한 원반 모양의 섬이라고 생각했다. 그는 이 세상의 모든 사물이 물로 이루어져 있다고 여겼다. 따라서 태양이나 달, 별도 물로 되어 있다고 보았다. 태양이나 달이 밝게 빛나는 이유는 이들이 물의 증기 상태로 구성되었기 때문이라고 생각했다.

비록 현재의 정확한 데이터와는 다르지만 탈레스는 태양의 궤도를 처음으로 규정했고, 1년을 365일로 나누는 것도 알고 있었다.

탈레스의 제자인 아낙시만드로스(Anaximandros, B.C.610~B.C.546)는 천문학 연구에 적절한 실험을 도입했다. 수직으로 세워 놓은 막대기의 그림자가 어떻게 움직이는가를 조사하여, 1년의 길이와 계절을 정확하게 알아냈다. 그는 태양, 달, 별 등은 불로 되어 있다고 믿었으며, 우리는 하늘의 돔(dome)에 뚫린 '움직이는 구멍'을 통하여 그 불을 보는 것으로 여겼다.

천구 모형의 등장_별이 있는 곳

정교수　이번에는 별이 있는 곳을 생각한 사람들의 이야기를 할 거야. 고대 그리스의 에우독소스가 별이 있는 곳을 최초로 언급했어. 그에 대해 먼저 알아보겠네.

에우독소스(Eudoxus of Cnidus, B.C.408~B.C.355)는 고대 그리스의 수학자이자 천문학자로 플라톤의 제자이다. 그의 이름 에우독소스(εὔδοξος)는 '영예로운' 또는 '평판이 좋은'을 의미한다. 고대 그리스어로 eu는 '좋은', doxa는 '의견, 신념, 평판'을 뜻한다.

에우독소스의 아버지 아이스키네스는 그리스 크니도스 출신으로 밤에 별을 보는 것을 좋아했다.

기원전 387년경, 23세의 나이에 에우독소스는 공부를 위해 아테네로 갔다. 그곳에서 몇 개월 동안 플라톤을 비롯한 여러 철학자의 강의를 들었다. 그는 매우 가난해 플라톤의 강의를 듣기 위해서는 매일 11km를 왕복해야 했다.

그 후 이집트로 간 에우독소스는 많은 제자를 모을 수 있었다. 기원전 368년경, 그는 제자들과 함께 아테네로 돌아왔다. 그리고 다시 크니도스로 돌아가서 천문대를 짓고 신학, 천문학, 기상학에 대한 책을 쓰고 강의를 계속했다.

에우독소스는 스승인 플라톤의 생각을 받아들여 지구가 우주의 중심이고 다른 천체들이 그 주위를 돈다고 생각했다. 그의 저서는 남아 있지 않으며, 그가 연구한 내용은 아리스토텔레스의 책을 통해 전해진다. 에우독소스는 동심구의 개념을 최초로 우주에 적용했다.

물리군 동심구가 뭐죠?

정교수 　같은 중심을 갖지만 반지름이 다른 구를 동심구라고 해.

　에우독소스는 행성의 일관되지 않은 움직임을 설명하고 천체의 움직임을 정확하게 계산하기 위해 동심구에 관한 아이디어를 내놓았다. 그는 27개의 동심구를 통해 우주를 이루는 각 천체들의 운동을 설명했다. 그는 태양을 제외한 모든 별이 지구를 중심으로 같은 거리에 있으며, 이 별들이 붙어 있는 구(천구)가 우주의 모든 동심구 중에서 최외각에 있다고 보았다. 이것이 바로 천구에 대한 최초의 생각이었다.
　그리스의 아리스토텔레스는 이 이론을 이어받았다. 그는 에우독소스와 비슷하게 동심구의 개념을 따랐지만, 태양과 달과 다섯 개의 행성 및 별들이 모두 붙어 있는 천구가 지구를 중심으로 원운동을 한다고 생각했다. 지구로부터 가까운 순서로 달, 수성, 금성, 태양, 화성,

아리스토텔레스의 천구 모형

목성, 토성이 원운동을 하며, 가장 먼 별들이 천구에 붙어 있다고 여겼다.

아리스토텔레스는 우주에 끝이 있고 그 끝은 천구라고 생각했다. 지구는 돌지 않고 단지 천구가 돌기 때문에 별자리가 움직이는 것으로 보았다.

물리군 지구가 움직이지 않는다고 생각하니까 별들이 천구에 붙어 회전한다고 믿었군요.

정교수 맞아. 또한 아리스토텔레스는 지구가 공 모양인 것을 최초로 확인한 사람이야.

물리군 지구가 공 모양이라고 처음 주장한 사람은 누구죠?

정교수 피타고라스 정리로 유명한 피타고라스라네.

물리군 그렇군요. 아리스토텔레스는 어떻게 지구가 공 모양이라는 것을 알았나요?

정교수 그는 월식 때 달을 가리는 어두운 부분이 지구의 그림자라고 생각했어. 그 그림자의 경계선이 원 모양이므로 지구가 둥그런 공 모양일 거라고 추측했지.

아리스토텔레스가 지구가 공 모양이라고 생각한 또 다른 근거가 있다. 수평선 너머로 배가 사라질 때 처음에는 선체가, 다음은 돛이, 마지막으로 돛대의 맨 꼭대기가 보이지 않게 된다. 이것은 지구가 공처럼 둥글지 않고 평평하다면 있을 수 없는 일이다. 아리스토텔레스

세상에서 가장 쉬운 과학 수업 별의 물리학

의 이러한 생각은 많은 이의 지지를 받아 당시 사람들은 지구가 둥글다는 것에 이의를 제기하지 않았다.

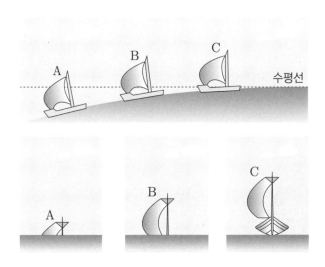

한편 아낙사고라스는 달이 태양 빛을 반사해 빛을 내는 것을 처음으로 알아냈다. 또한 태양이 신이 아니라 백열 상태로 빛나고 있는 돌멩이에 불과하다고 주장했다. 이 언급 때문에 그는 신에 대한 불경죄로 재판을 받았다. 친구 사이였던 아테네의 지도자 페리클레스가 변호 연설을 해주었지만, 결국 아낙사고라스는 아테네에서 추방되어 람프사코스에 갇혀 지내게 되었다.

원자론으로 유명한 데모크리토스는 천구 바깥에 물질이 없다고 주장했다. 그는 천구 밖은 아무것도 없는 진공 상태라고 생각했다.

아리스타르코스의 관측_지구와 달, 지구와 태양 사이의 거리

정교수 지구와 달, 지구와 태양 사이의 거리를 처음 계산한 사람이 있어.

물리군 그게 누구예요?

정교수 고대 그리스의 아리스타르코스야.

아리스타르코스(Aristarchus of Samos, B.C. 310~B.C. 230)

아리스타르코스는 삼각비[1]를 이용해 지구와 달 사이의 거리와 지구와 태양 사이의 거리의 비를 구했다네.

물리군 어떤 방법으로요?

정교수 그가 사용한 방법은 간단해.

─────────────

1) 고대 그리스 시대에 sin, cos 같은 기호는 없었지만 닮음으로부터 삼각비 개념은 알고 있었다.

아리스타르코스는 태양과 반달이 하늘에 동시에 나타날 때 달을 유심히 관찰했다. (상현달은 정오를 넘긴 오후에, 하현달은 오전에 해와 함께 볼 수 있다.) 달이 빛나는 것은 태양 빛을 반사하기 때문이라는 사실로부터, 둥근 달의 반을 나누는 경계선과 수직 방향에 태양이 있음을 알 수 있다. 이를 바라보는 관측자가 지구에 있는 것을 고려할 때 지구와 달, 태양을 연결하면 직각삼각형이 된다.

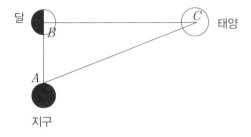

이때 $\angle BAC = \theta$로 놓으면 각 B가 직각이므로

$$\cos \theta = \frac{\overline{AB}}{\overline{AC}}$$

이다. $\overline{AB}$ 는 지구와 달 사이의 거리이고, $\overline{AC}$ 는 지구와 태양 사이의 거리이므로 θ를 알면 두 거리의 비를 알 수 있다. 아리스타르코스의 관측에 의하면 $\theta = 87°$였다. 이 값을 위 식에 대입하면

$$\overline{AC} \cong 19 \times \overline{AB}$$

가 된다.

물리군　생각보다 단순한 방법이군요.

정교수　하지만 아리스타르코스의 관측은 정확하지 않았어. 실제로 지구와 태양 사이의 거리는 지구와 달 사이의 거리의 약 400배 정도라네.

지구의 반지름을 계산한 에라토스테네스_태양 광선과 그림자를 이용해

정교수　지구가 공 모양이라는 것이 알려진 이후 지구의 반지름을 측정하겠다고 도전한 학자는 바로 고대 그리스의 에라토스테네스야.

에라토스테네스
(Eratosthenes of Cyrene, B.C.276~B.C.195?)

물리군　에라토스테네스는 어떻게 지구의 반지름을 알아냈나요?

정교수　그 원리를 지금부터 설명해 볼게.

기원전 240년경 알렉산드리아의 도서관장이었던 에라토스테네스는 지구의 반지름을 최초로 계산했다.

에라토스테네스는 알렉산드리아에서 남동쪽으로 925km 떨어진 시에네(현재의 아스완)에서 하짓날 정오에 태양 광선이 수직으로 비추고, 알렉산드리아에서는 태양 광선이 7.2°만큼 비스듬히 비춘다는 사실을 이용하여 지구의 둘레를 측정했다.

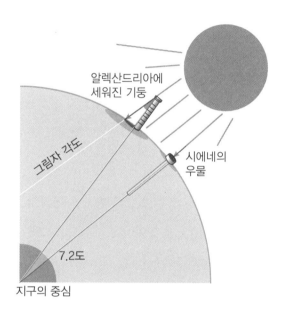

이제 에라토스테네스가 어떻게 지구의 반지름을 측정했는지 수학적으로 알아보자.

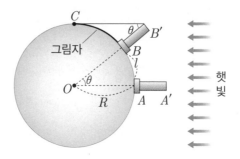

A를 시에네, B를 알렉산드리아라고 하자. 에라토스테네스는 다음과 같은 두 가지 중요한 가정을 세웠다.

첫째, 지구는 완전한 공 모양이다.
둘째, 지구로 들어오는 태양 광선은 평행하다.

막대 AA'은 그림자가 생기지 않도록 햇빛과 나란히 세우고, 막대 BB'은 막대 AA'과 같은 경도에 놓이도록 세운다. $\angle BB'C = \theta$라고 하면 $\angle BOA$는 $\angle BB'C$와 엇각으로 같다. 그러므로 $\angle BOA = \theta$이다.

이때 부채꼴 BOA의 중심각 $\angle BOA$에 대응하는 부채꼴의 호의 길이는 호 AB의 길이인 l이다. 부채꼴의 호의 길이는 중심각에 비례하며, 원둘레의 길이에 대한 중심각은 360°이다. 따라서 지구의 반지름을 R라고 하면 다음과 같은 비례식을 세울 수 있다.

$$\theta : l = 360° : 2 \times \pi \times R$$

여기서 R를 구하면

$$R = \frac{180° \times l}{\pi \times \theta}$$

이다. 이 방법으로 에라토스테네스는 지구의 반지름이 약 6275km라는 것을 알아냈다.[2]

별의 밝기를 연구한 히파르코스_별의 밝기에 등급을 매기다

정교수   이번에는 별의 밝기에 등급을 매긴 이야기를 해 보겠네. 이 일을 한 사람은 고대 그리스의 히파르코스일세.

히파르코스(Hipparchus, B.C.190~B.C.120)

2] 현재 정확한 지구 반지름은 약 6367km이다.

히파르코스는 로도스섬에 관측소를 만들어 별을 관측했다. 훗날 로마 시대에 사용된 1022개의 별들 중 850개를 그가 발견했다고 전해진다. 그는 별을 밝기에 따라 1등성부터 6등성까지로 구분했는데, 가장 밝은 별을 1등성이라 하고 눈에 겨우 보이는 별을 6등성으로 정의했다.

히파르코스 이후, 거의 2천 년 동안이나 1등성이 2등성보다 얼마나 더 밝은 것인지 제대로 알려지지 않았다. 그러다 1865년 영국의 천문학자 포그슨(Norman Robert Pogson, 1829~1891)이 처음으로 히파르코스가 정한 1등급의 별이 6등급의 별에 비해 약 100배 밝은 것을 알아냈다. 즉, 다섯 등급이 100배의 밝기 차이를 가지므로 한 등급은 약 2.512배의 밝기 차이가 난다.

세상에서 가장 쉬운 과학 수업 별의 물리학

두 번째 만남

•

만유인력과 별의 거리

만유인력의 발견_질량을 가진 두 물체가 서로 끌어당기는 힘

정교수　별의 물리학으로 들어가려면 먼저 만유인력을 이해할 필요가 있다네. 만유인력이 뭔지는 알고 있지?

물리군　질량을 가진 두 물체 사이에 서로를 끌어당기는 힘이라고 배웠어요.

정교수　맞아. 그럼 만유인력의 역사적 배경부터 한번 살펴볼게.

　만유인력의 법칙을 알아낸 사람은 영국의 과학자 뉴턴이다. 뉴턴은 천체들이 영원한 운동을 하는 이유에 대해 고민했다. 그리고는 태양을 중심으로 공전하는 행성(planet)과 태양 사이에 만유인력이라는 힘이 존재하는 것을 발견했다.

　이와 더불어 만유인력이 태양의 질량과 행성의 질량의 곱에 비례하고, 태양과 행성 사이의 거리의 제곱에 반비례함을 알아냈다. 이 힘은 서로를 잡아당기는 인력인데 여기서 뉴턴은 새로운 고민에 빠졌다. 천체들 사이에 서로를 잡아당기는 힘이 작용하는데 왜 두 천체는 서로 달라붙지 않을까? 뉴턴은 그 이유가 행성이 태양 주위를 공전하기 때문임을 깨달았다. 그는 이 내용을 저서 《프린키피아》에서 자세하게 다루었다.

　뉴턴은 먼저 행성의 운동에 대한 케플러의 연구 결과를 들여다보았다. 케플러는 태양 주위를 행성이 공전할 때, 행성의 운동은 태양을 한 초점으로 하는 타원운동인 것을 밝혀냈다.

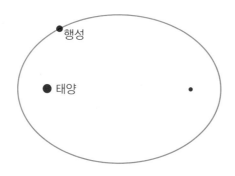

또한 케플러는 행성이 한 번 공전하는 데 걸리는 시간(공전주기)을 T, 타원의 긴반지름을 a라고 하면

$$T^2 = Ka^3 \tag{2-1-1}$$

이 성립함을 알아냈다. 즉, 공전주기의 제곱은 긴반지름의 세제곱에 비례한다는 것이다. 뉴턴은 이 케플러의 법칙을 수학적으로 완벽하게 증명해 《프린키피아》에 수록했다.

뉴턴은 질량 M인 물체와 질량 m인 물체가 거리 r만큼 떨어져 있을 때, 두 물체 사이의 만유인력의 크기 F는

$$F = \frac{GMm}{r^2} \tag{2-1-2}$$

인 것을 확인했다. 여기서 G는 중력 상수 또는 뉴턴 상수라고 하며,

$$G = 6.674 \times 10^{-11} \mathrm{m}^3 \cdot \mathrm{kg}^{-1} \cdot \mathrm{s}^{-2} \tag{2-1-3}$$

으로 나타낸다.

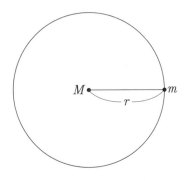

물리군 《프린키피아》는 읽은 적이 없어서 잘 모르겠어요.

정교수 뉴턴의 케플러 법칙 증명은 여러 단계를 거친다네. 자세한 내용은 나중에 일반상대성이론을 다루는 책에서 강의하도록 하지.

　타원의 특별한 경우인 원을 생각하자. 즉, 긴반지름과 짧은반지름의 길이가 같은 타원이 원이다. 이제 질량 m인 행성이 질량 M인 태양 주위로 반지름 r인 원운동을 하는 경우를 보자. 원운동을 하기 위해서는 행성이 구심력을 받아야 한다. 그 구심력은 태양이 행성에 작용하는 인력이다. 태양이 행성에 작용하는 인력(만유인력)의 크기를 F라고 하면

　$F = (행성이 원운동을 하기 위한 구심력)$

세상에서 가장 쉬운 과학 수업 별의 물리학

이다.

원운동을 하기 위해 왜 구심력이 필요한지 알아보자. 양동이에 물을 가득 담고 거꾸로 뒤집으면 양동이 속의 물이 바닥에 떨어진다. 이것은 지구가 물을 잡아당기는 만유인력 때문이다. 하지만 양동이를 빙글빙글 돌리면 양동이가 뒤집어져도 그 속의 물은 바닥에 떨어지지 않는다. 뉴턴은 달이 지구 주위를 돌면서도 지구로 떨어지지 않는 이유는 달이 지구를 중심으로 원운동을 하기 때문임을 발견했다.

원운동하는 물체는 원운동을 하게 하는 힘이 있어야 하는데, 그 힘을 뉴턴은 구심력이라고 불렀다. 달이 지구 주위를 원운동하게 하는 힘은 만유인력이다. 그러므로 달과 지구 사이의 만유인력이 바로 구심력이다.

행성이 원운동을 하기 위한 구심력은

$$m\frac{v^2}{r} \qquad\qquad (2-1-4)$$

이다. 여기서 v는 공전속도이다. 행성의 공전주기를 T라고 하면

$$v = \frac{2\pi r}{T} \qquad\qquad (2-1-5)$$

가 된다. 따라서

$$
\begin{aligned}
(\text{만유인력의 크기}) &= m\frac{v^2}{r} \\
&= \frac{m}{r}\left(\frac{2\pi r}{T}\right)^2
\end{aligned}
$$

이다. 한편 케플러 법칙 (2-1-1)로부터

$$T^2 = Kr^3$$

이므로

$$(만유인력의 크기) = \frac{4\pi^2 m}{K}\left(\frac{1}{r^2}\right) \tag{2-1-6}$$

이 되어, 만유인력의 크기가 거리의 제곱에 반비례함을 알 수 있다.

중력가속도의 계산_중력 때문에 생긴 가속도

정교수   이번에는 중력가속도를 계산해 보겠네. 질량 m인 물체가 질량 M인 물체 때문에 만유인력을 받는 경우를 생각해 보세. 만유인력은 다른 말로 중력이라고도 해.[3] 질량 m인 물체는 힘(중력)을 받으니까 가속도가 생기지. 이렇게 중력 때문에 생긴 가속도를 중력가속도라 하고 g로 쓴다네.

질량 m인 물체가 질량 M인 물체와 거리 r만큼 떨어져 있을 때, 질량 m인 물체가 받는 힘은

3] 과거에는 만유인력과 중력을 구분해서 정의했지만 지금은 만유인력 대신 주로 중력이라는 용어를 사용한다.

$$F = mg$$

이므로 중력가속도는

$$g = \frac{GM}{r^2} \qquad (2\text{-}2\text{-}1)$$

이 된다. 즉, 그 위치에서의 중력가속도를 알면 물체가 받는 힘은 물체의 질량과 중력가속도의 곱으로 구할 수 있다.

이제 식 (2-2-1)을 다음과 같이 나타내자.

$$g \times 4\pi r^2 = 4\pi GM \qquad (2\text{-}2\text{-}2)$$

반지름 r인 구를 생각하여 그 중심에 질량 M인 물체가 있다고 하자. 이때 구 표면 위의 모든 점에서 중력가속도의 크기는 같다. 그러므로 식 (2-2-2)는 다음과 같이 나타낼 수 있다.

(반지름 r인 구 표면에서의 중력가속도) × (구의 표면적)

= $4\pi G$ × (구 속에 포함된 질량)

즉, 구의 표면적을 A, 표면에서의 중력가속도를 g라고 하면

$$g \times A = 4\pi G \times (구 \ 속에 \ 포함된 \ 질량) \qquad (2\text{-}2\text{-}3)$$

이다.

물리군 식 (2-2-3)은 뉴턴이 알아냈나요?

정교수 그건 뉴턴이 죽은 후 프랑스의 뉴턴으로 불린 라플라스가 연구한 내용이야. 식 (2-2-3)을 사용하는 예를 하나 들어 볼게.

라플라스
(Pierre–Simon Marquis de Laplace, 1749~1827)

반지름이 R인 공 모양의 물체를 생각하자. 이 물체의 질량은 M이고 밀도는 균일하다고 하자. 밀도는 질량을 부피로 나눈 값이므로 이 물체의 밀도를 ρ라고 하면

$$\rho = \frac{M}{\frac{4\pi}{3}R^3} \qquad (2-2-4)$$

이 된다. 이때 공 내부에서의 중력가속도를 구해 보자. 다음 그림과 같이 반지름이 R인 구 속에 반지름이 r인 동심구를 그리자.

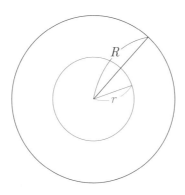

반지름이 r인 구 표면에서의 중력가속도는 모두 같다. 이 중력가속도를 $g(r)$라고 하면 식 (2-2-3)으로부터

$$g(r) \times 4\pi r^2 = 4\pi G \times M(r) \qquad (2\text{-}2\text{-}5)$$

가 된다. 여기서 $M(r)$는 반지름이 r인 구 속에 포함된 질량을 말한다.

물리군 $M(r)$는 어떻게 구하죠?

정교수 밀도에 부피를 곱하면 돼. 그러니까

$$M(r) = \rho \times \frac{4\pi}{3} r^3 = M\left(\frac{r}{R}\right)^3 \qquad (2\text{-}2\text{-}6)$$

이지. 이것을 식 (2-2-5)에 대입하면

$$g(r) = \left(\frac{GM}{R^3}\right) r \qquad (2\text{-}2\text{-}7)$$

가 되네. 그러니까 구의 내부에서 중력가속도는 구의 중심에서 멀어질수록 점점 커지는 걸세.

물리군 식 (2-2-7)은 중요한가요?

정교수 앞으로 많이 사용될 거야.

중력에 대한 퍼텐셜 에너지_미분과 적분을 이용하여

정교수 이제 중력에 대응하는 퍼텐셜 에너지를 알아보기로 하세.

3차원에서 질량 m인 물체에 힘 $\vec{F}$가 작용하면 뉴턴의 운동법칙에 의해

$$\vec{F} = m\vec{a} \tag{2-3-1}$$

이다. 여기서 가속도벡터 $\vec{a}$는 속도벡터 $\vec{v}$를 시간으로 미분한 것이므로

$$\vec{a} = \frac{d\vec{v}}{dt} \tag{2-3-2}$$

이다.

물체에 힘이 작용해 물체가 어떤 방향으로 일정한 거리만큼 이동한 경우를 생각하자.

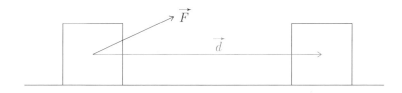

그림에서 $\vec{d}$의 크기는 이동한 거리를, 방향은 이동 방향을 나타낸다. 이때 이 힘이 한 일을 물리학자들은 W로 쓰고, 다음과 같이 두 벡터의 내적으로 정의한다.

$$W = \vec{F} \cdot \vec{d}$$

(2-3-3)

이제 힘 $\vec{F}$가 질량 m인 물체에 작용해 다음 그림의 A에서 B로 이동하는 경우를 보자.

이 곡선 경로를 무한히 잘게 등분한 작은 토막을 생각하자. 너무나 작은 토막이므로 그 길이는 거의 0이다. 그중 임의의 한 토막이 P에서 Q로 가는 경로라고 하자. 매우 작게 등분했기 때문에 P에서 Q로 가는 곡선과 P에서 Q로 가는 직선은 거의 같다.

다음 그림과 같이 원점 O를 정하여 점 P의 위치벡터를 $\vec{r}$라 하고,

P에서 Q로 향하는 벡터를 $\vec{dr}$라고 하자. 이 벡터는 길이 요소 벡터로 부르는데, 각각의 토막에서 서로 다른 방향을 향하지만 그 크기는 거의 0에 가까워진다.

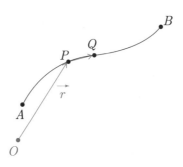

이때 점 P에서 물체에 작용하는 힘을 $\vec{F}$라고 하자.

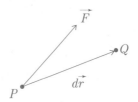

P와 Q 사이의 거리가 거의 0에 가까울 정도로 작기 때문에, 물체가 P에서 Q로 갈 때 힘은 달라지지 않는다. 그러므로 이 힘을 받아 물체가 P에서 Q로 갈 때 물체가 한 일을 dW라고 하면

$$dW = \vec{F} \cdot d\vec{r} \qquad (2\text{-}3\text{-}4)$$

이다. 그러므로 힘 $\vec{F}$ 가 작용해 질량 m인 물체가 A에서 B로 이동할 때 이 힘이 한 일 W_{AB}는

$$W_{AB} = \int_A^B \vec{F} \cdot d\vec{r} \qquad (2\text{-}3\text{-}5)$$

가 된다. 이 식은

$$W_{AB} = \int_A^B m\frac{d\vec{v}}{dt} \cdot d\vec{r}$$

$$= \int_A^B m\frac{d\vec{v}}{dt} \cdot \frac{d\vec{r}}{dt} dt$$

로 쓸 수 있다. 한편

$$\vec{v} = \frac{d\vec{r}}{dt}$$

이므로

$$W_{AB} = \int_A^B m\frac{d\vec{v}}{dt} \cdot \vec{v} dt \qquad (2\text{-}3\text{-}6)$$

가 된다. 이제 단위벡터를 이용해서 벡터를 성분으로 나타내자.

$$\vec{v} = v_x \hat{i} + v_y \hat{j} + v_z \hat{k} \tag{2-3-7}$$

라고 하면

$$\frac{d\vec{v}}{dt} = \frac{dv_x}{dt} \hat{i} + \frac{dv_y}{dt} \hat{j} + \frac{dv_z}{dt} \hat{k} \tag{2-3-8}$$

이므로

$$\begin{aligned} W_{AB} &= \int_A^B m \left(v_x \frac{dv_x}{dt} + v_y \frac{dv_y}{dt} + v_z \frac{dv_z}{dt} \right) dt \\ &= \int_A^B \frac{m}{2} \frac{d}{dt} \left(v_x^2 + v_y^2 + v_z^2 \right) dt \end{aligned} \tag{2-3-9}$$

가 된다. 따라서 물체가 3차원에서 운동을 할 때 운동 에너지를 K라 하고

$$K = \frac{1}{2} m \left(v_x^2 + v_y^2 + v_z^2 \right) = \frac{1}{2} m \vec{v} \cdot \vec{v} \tag{2-3-10}$$

로 정의하면

$$W_{AB} = K_B - K_A \tag{2-3-11}$$

이다. 이때 K_B, K_A는 각각 점 B, 점 A에서의 운동 에너지이다.

주어진 힘 $\vec{F}$에 대해 다음 조건을 만족하는 함수 $U(x, y, z)$가 존재할 때, 이 함수를 퍼텐셜 에너지, 이 힘을 보존력이라고 부른다.

$$\vec{F} = -\vec{\nabla}U \tag{2-3-12}$$

여기서 $\vec{\nabla}$ 를 벡터미분연산자라 하고

$$\vec{\nabla} = \hat{i}\frac{\partial}{\partial x} + \hat{j}\frac{\partial}{\partial y} + \hat{k}\frac{\partial}{\partial z} \tag{2-3-13}$$

로 정의한다. 이때 $\frac{\partial}{\partial x}$, $\frac{\partial}{\partial y}$, $\frac{\partial}{\partial z}$ 는 각각 x, y, z에 대한 편미분이라고 부른다.

따라서 힘 $\vec{F}$를

$$\vec{F} = F_x\hat{i} + F_y\hat{j} + F_z\hat{k} \tag{2-3-14}$$

로 놓으면

$$F_x = -\frac{\partial U}{\partial x}$$

$$F_y = -\frac{\partial U}{\partial y}$$

$$F_z = -\frac{\partial U}{\partial z} \tag{2-3-15}$$

가 된다. 이때

$$\vec{r} = x\hat{i} + y\hat{j} + z\hat{k} \tag{2-3-16}$$

라고 하면

$$\vec{dr} = dx\hat{i} + dy\hat{j} + dz\hat{k} \qquad (2\text{-}3\text{-}17)$$

이므로

$$W_{AB} = \int_A^B (F_x dx + F_y dy + F_z dz)$$

$$= -\int_A^B \left(\frac{\partial U}{\partial x} dx + \frac{\partial U}{\partial y} dy + \frac{\partial U}{\partial z} dz \right) \qquad (2\text{-}3\text{-}18)$$

이다. 전미분의 정의

$$dU = \frac{\partial U}{\partial x} dx + \frac{\partial U}{\partial y} dy + \frac{\partial U}{\partial z} dz$$

를 이용하면

$$W_{AB} = -\int_A^B dU = -(U_B - U_A) \qquad (2\text{-}3\text{-}19)$$

임을 알 수 있다. 여기서 U_B, U_A는 각각 점 B, 점 A에서의 퍼텐셜 에너지이다. 식 (2-3-11), (2-3-19)로부터

$$W_{AB} = K_B - K_A = -(U_B - U_A) \qquad (2\text{-}3\text{-}20)$$

가 되어, 임의의 두 지점에서 $K + U$는 같아진다. 임의의 두 지점에서 같은 것은 운동 중인 물체에 대해 모든 점에서 이 값이 같음을 뜻한

세상에서 가장 쉬운 과학 수업 별의 물리학

다. 이 양을 역학적 에너지라 하고 E로 쓴다.

$$E = K + U \qquad\qquad (2\text{-}3\text{-}21)$$

이것을 역학적 에너지 보존법칙이라고 부른다.

물리군   전미분에 대해 좀 더 알고 싶어요.

정교수   고등학교에서는 변수가 x 한 개인 함수를 배우지. 이것을 일변수함수라고 불러.

물리군 $y = f(x)$를 말하나요?

정교수   맞아. 그런데 지금 우리가 다루는 퍼텐셜 에너지는 세 개의 변수 x, y, z를 가진 함수야. 이렇게 변수가 세 개인 함수를 삼변수함수, 변수가 두 개인 함수를 이변수함수라고 한다네.

예를 들어 이변수함수 $f(x, y)$를 보자. 이 함수의 전미분은 df로 쓰고 다음과 같이 정의한다.

$$df = f_x dx + f_y dy \qquad\qquad (2\text{-}3\text{-}22)$$

여기서 우리는 편미분에 대해 간단한 기호

$$\frac{\partial f}{\partial x} = f_x$$

$$\frac{\partial f}{\partial y} = f_y$$

를 사용했다. 마찬가지로 삼변수함수 $f(x, y, z)$에 대한 전미분은 다음과 같다.

$$df = f_x dx + f_y dy + f_z dz \qquad (2\text{-}3\text{-}23)$$

일변수함수에서는 $f(x) = $ (상수)이면 $f'(x) = 0$이고 그 역도 성립한다. 이변수함수 $f(x, y)$에서 $f(x, y) = $ (상수)이면 $f_x = 0$이고 $f_y = 0$이 된다. 그럼 $f_x = 0$이면 f는 상수일까? 그렇지는 않다. 예를 들어 $f(x, y) = y$를 보면 $f_x = 0$이지만 f가 상수는 아니다. 그러므로 이변수함수에서 $f(x, y) = $ (상수)라는 결과를 만들려면 $f_x = 0$이고 $f_y = 0$이어야 한다. 즉,

$$df = 0$$

의 조건을 요구해야 한다.

이제 전미분과 적분의 관계를 알아보자. 일변수함수 $f(x)$의 경우 전미분은

$$df = \frac{df}{dx} dx$$

이므로 적분은 다음과 같이 정의한다.

$$\int_0^x df = f(x) - f(0) \qquad (2\text{-}3\text{-}24)$$

이것은 수직선상에서 0부터 x까지의 적분이다.

세상에서 가장 쉬운 과학 수업 별의 물리학

이변수함수의 경우는 다음과 같이 2차원 평면에서 점 (0, 0)부터 점 (x, y)까지의 적분을 고려해야 한다.

$$\int_{(0,0)}^{(x,y)} df = f(x, y) - f(0, 0) \qquad (2\text{-}3\text{-}25)$$

물리군 일변수함수와 비교하니까 전미분이 이해가 되네요. 그럼 중력에 대한 퍼텐셜 에너지는 뭐죠?

정교수 질량 m인 물체가 질량 M인 물체로 인해 중력을 받는 경우를 생각해 보게.

질량 M인 물체를 원점에 놓고 질량 m인 물체의 위치벡터를 $\vec{r}$ 라고 하자.

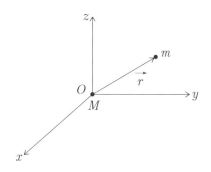

질량 m인 물체가 받는 중력 $\vec{F}$ 는 인력이니까 다음 그림과 같다.

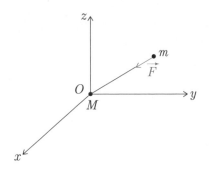

방향이 $\vec{r}$ 와 같으면서 크기가 1인 벡터는 항상 만들 수 있다. 이 벡터는 $\vec{r}$ 를 크기로 나누면 구해진다. 이것을 $\hat{r}$ 로 쓰고 'hat r'라고 읽는다. 즉,

$$\hat{r} = \frac{\vec{r}}{r} \tag{2-3-26}$$

이다. 여기서

$$r = |\vec{r}| = \sqrt{x^2 + y^2 + z^2} \tag{2-3-27}$$

이 된다. 그런데 질량 m인 물체가 받는 중력은 $\hat{r}$와 반대 방향이므로

$$\vec{F} = -\frac{GMm}{r^2}\hat{r} \tag{2-3-28}$$

이다. 이 식은 다음과 같이 쓸 수 있다.

세상에서 가장 쉬운 과학 수업 별의 물리학

$$\vec{F} = -\frac{GMm}{r^3}\vec{r} \qquad (2\text{-}3\text{-}29)$$

따라서 중력에 대한 퍼텐셜 에너지 U는

$$-\vec{\nabla}U = \vec{F} = -\frac{GMm}{r^3}\vec{r} \qquad (2\text{-}3\text{-}30)$$

로부터 구할 수 있다.

물리군 좀 복잡한데요.

정교수 각각의 성분을 비교하면 돼. x, y, z성분을 비교하면

$$\frac{\partial U}{\partial x} = GMm\frac{x}{r^3}$$

$$\frac{\partial U}{\partial y} = GMm\frac{y}{r^3}$$

$$\frac{\partial U}{\partial z} = GMm\frac{z}{r^3} \qquad (2\text{-}3\text{-}31)$$

라네. 여기서 U를 구하면 되지.

물리군 어떻게 구하죠?

정교수 차근차근 설명해 주겠네.

r를 x로 편미분하면 다음과 같다.

$$\frac{\partial r}{\partial x} = \frac{\partial}{\partial x}\sqrt{x^2 + y^2 + z^2}$$

$$= \frac{1}{2\sqrt{x^2 + y^2 + z^2}} \times 2x$$

$$= \frac{x}{r}$$

마찬가지로

$$\frac{\partial r}{\partial y} = \frac{y}{r}$$

$$\frac{\partial r}{\partial z} = \frac{z}{r}$$

가 된다. 이제 식 (2-3-31)을 만족하는 U가 r만의 함수라고 가정하자. 이때

$$\frac{\partial U(r)}{\partial x} = U'(r)\frac{\partial r}{\partial x} = U'(r)\frac{x}{r}$$

이고, 같은 방법으로

$$\frac{\partial U(r)}{\partial y} = U'(r)\frac{\partial r}{\partial y} = U'(r)\frac{y}{r}$$

$$\frac{\partial U(r)}{\partial z} = U'(r)\frac{\partial r}{\partial z} = U'(r)\frac{z}{r}$$

세상에서 가장 쉬운 과학 수업 별의 물리학

가 된다. 여기서 $U'(r) = \dfrac{dU}{dr}$ 이다. 따라서 식 (2-3-31)은

$$U'(r)\frac{x}{r} = GMm\frac{x}{r^3}$$

$$U'(r)\frac{y}{r} = GMm\frac{y}{r^3}$$

$$U'(r)\frac{z}{r} = GMm\frac{z}{r^3} \tag{2-3-32}$$

로 쓸 수 있다. 이 식은

$$U'(r) = GMm\frac{1}{r^2} \tag{2-3-33}$$

이 되므로

$$U(r) = -\frac{GMm}{r} \tag{2-3-34}$$

이다. 이것이 바로 질량 M인 물체로부터 거리 r만큼 떨어진 곳에 있는 질량 m인 물체가 받는 중력에 대한 퍼텐셜 에너지이다.

비리얼 정리_클라우지우스의 증명

정교수　이번에는 비리얼(virial) 정리를 소개할게.

물리군　비리얼이 뭔가요?

정교수　비리얼은 힘을 나타내는 라틴어 vis를 이용해 물리학자 클라우지우스가 만든 단어일세. 그가 1870년에 비리얼 정리를 증명했거든.

클라우지우스
(Rudolf Julius Emanuel Clausius, 1822~1888)

　클라우지우스는 주기적인 운동을 생각했다. 물체가 주기적인 운동을 할 때 그 주기를 T라고 하면, 물체의 위치 $\vec{r}$와 속도 $\vec{v}$는 주기적이므로

$$\vec{r}(t) = \vec{r}(t+T)$$

$$\vec{v}(t) = \vec{v}(t+T) \tag{2-4-1}$$

이다. 이 식에 $t = 0$을 대입하면

$$\vec{r}(T) = \vec{r}(0)$$

$$\vec{v}(T) = \vec{v}(0) \tag{2-4-2}$$

이 된다.

질량 m인 물체에 대해 다음과 같은 양을 생각해 보자.

$$G = m\vec{v} \cdot \vec{r} \tag{2-4-3}$$

$$I = m\vec{r} \cdot \vec{r} \tag{2-4-4}$$

식 (2-4-4)를 시간으로 미분하면

$$\frac{dI}{dt} = 2m\vec{r} \cdot \frac{d\vec{r}}{dt} = 2m\vec{r} \cdot \vec{v}$$

이므로

$$\frac{dI}{dt} = 2G \tag{2-4-5}$$

가 된다.

이제 G를 시간으로 미분하자.

$$\frac{dG}{dt} = m\frac{d\vec{v}}{dt} \cdot \vec{r} + m\vec{v} \cdot \vec{v}$$

$$= \vec{F} \cdot \vec{r} + m\vec{v} \cdot \vec{v}$$

$$= \vec{F} \cdot \vec{r} + 2K \qquad (2\text{-}4\text{-}6)$$

어떤 양 X의 시간 평균을 다음과 같이 정의하자.

$$<X> = \frac{1}{T}\int_0^T X(t)\,dt \qquad (2\text{-}4\text{-}7)$$

식 (2-4-6)에서 양변의 시간 평균을 구하면

$$<\frac{dG}{dt}> = \frac{1}{T}\int_0^T \frac{dG}{dt}dt$$

$$= \frac{1}{T}\{G(T) - G(0)\}$$

이다. G는 주기함수들의 내적으로 주기함수이다. 따라서

$$G(T) = G(0)$$

이므로

$$<\frac{dG}{dt}> = 0 \qquad (2\text{-}4\text{-}8)$$

이 되어 식 (2-4-6)으로부터

$$< \vec{F} \cdot \vec{r} > + 2 < K > = 0 \qquad (2\text{-}4\text{-}9)$$

임을 알 수 있다. 보존력에 대해 위 식은

$$-< \vec{\nabla U} \cdot \vec{r} > + 2 < K > = 0 \qquad (2\text{-}4\text{-}10)$$

이 된다.

이제 퍼텐셜 에너지가 다음 꼴로 주어지는 경우를 생각하자.

$$U = Cr^{n} \qquad (2\text{-}4\text{-}11)$$

이때

$$\frac{\partial U}{\partial x} = Cnr^{n-1} \frac{\partial r}{\partial x} = Cnr^{n-1} \frac{x}{r} = Cnr^{n-2} x$$

$$\frac{\partial U}{\partial y} = Cnr^{n-1} \frac{\partial r}{\partial y} = Cnr^{n-1} \frac{y}{r} = Cnr^{n-2} y$$

$$\frac{\partial U}{\partial z} = Cnr^{n-1} \frac{\partial r}{\partial z} = Cnr^{n-1} \frac{z}{r} = Cnr^{n-2} z \qquad (2\text{-}4\text{-}12)$$

이다. 그러므로 다음과 같다.

$$\vec{\nabla} U \cdot \vec{r} = x\frac{\partial U}{\partial x} + y\frac{\partial U}{\partial y} + z\frac{\partial U}{\partial z}$$

$$= Cnr^{n-2}(x^2 + y^2 + z^2)$$

$$= Cnr^n$$

$$= nU \qquad\qquad (2\text{-}4\text{-}13)$$

이로부터 식 (2-4-10)은

$$<K> = \frac{n}{2}<U> \qquad\qquad (2\text{-}4\text{-}14)$$

가 된다. 이것을 비리얼 정리라고 부른다.

중력에 대한 비리얼 정리를 찾아보자. 중력의 경우 퍼텐셜 에너지는

$$U = Cr^{-1}$$

의 꼴이므로 비리얼 정리에 의해

$$<K> = -\frac{1}{2}<U> \qquad\qquad (2\text{-}4\text{-}15)$$

임을 알 수 있다.

세상에서 가장 쉬운 과학 수업 별의 물리학

광행차의 발견_지구 공전의 증거

정교수   이제 지구의 공전 증거를 찾은 과학자 브래들리의 연구를 알아볼게.

브래들리(James Bradley, 1692~1762)

브래들리는 1692년 영국 셔본에서 태어났다. 그는 1711년 옥스퍼드의 베일리얼 대학에 입학하여 1714년과 1717년에 각각 문학 학사와 석사 학위를 받았다. 그의 초기 관측은 그의 삼촌이자 천문학자인 파운드(James Pound)의 도움으로 이루어졌고, 이 관측 결과로 그는 1718년 왕립학회 회원으로 선출되었다. 이후 1721년 옥스퍼드의 새빌리언 천문학자 자리를 시작으로 1742년에는 핼리의 뒤를 이어 왕립 천문학자로 임명되었다.

여기서 잠깐 영국 최고의 천문대인 그리니치 천문대 이야기로 넘어가자. 그리니치 천문대는 1675년 8월 10일에 건립되었으며, 당시의 이름은 왕립 그리니치 천문대(Royal Greenwich Observatory)였다.

그리니치 천문대
(출처: ChrisO/Wikimedia Commons)

찰스 2세는 천문대를 설립할 때 존 플램스티드를 초대 천문대 대
장으로 임명하면서 왕실 천문관이라
는 호칭을 만들기도 했다. 현재 그리
니치 천문대는 런던 그리니치의 그리
니치 공원에 위치해 있다.

플램스티드는 영국 더비셔주의 작
은 마을 덴비에서 태어났다. 그는 더
비의 한 학교에서 1662년까지 라틴어
와 역사를 공부하였다. 이후 교장의
추천을 받아 케임브리지 대학의 지저

플램스티드(John Flamsteed, 1646
~1719, 사진 출처: Godfrey Kneller/
Wikimedia Commons)

세상에서 가장 쉬운 과학 수업 별의 물리학

스 칼리지에 입학할 기회를 얻었으나, 만성적인 건강 문제로 몇 년간 진학을 미루었다.

그는 집에 머무르며 아버지의 사업을 도왔고, 이때 산술과 분수 계산법 등을 배우고 수학과 천문학에 호기심을 보였다. 1662년 7월에는 13세기의 천문학자 요하네스 데 사크로보스코의 저서《천구론 (De sphaera mundi)》을 읽고 크게 감명받았다는 사실을 그의 일기로부터 알 수 있다.

1662년 9월 12일에는 생애 처음으로 부분일식을 관측하였다. 1663년 초에는 팔레(Thomas Fale)의 저서《Horologiographia: The Art of Dialling》을 읽고 해시계에 주목했고, 스트리트(Thomas Street)의《Astronomia Carolina》, 이른바 '천체의 운동에 관한 새로운 이론'을 공부하였다. 또한 천문학에 관심을 가진 지역 유지들과 폭넓게 사귀었는데, 그들의 서재를 드나들며 제러마이아 호록스의 천문표를 수록한 서적들을 접했다. 훗날 그는 호록스의 이론을 개선하여 달의 위치를 계산했다.

19세가 되던 1665년 10월, 플램스티드는 〈수학 에세이〉라는 제목의 첫 번째 논문을 작성하였다. 여기에는 천문 관측용 사분의를 설계하고 제작, 사용하는 방법과 더비 지역에서의 관측 결과를 다루고 있다.

플램스티드는 1670년 9월에야 케임브리지 대학 지저스 칼리지에 입학했다. 그는 케임브리지에 주로 거주하지는 않았지만 1674년에는 뉴턴의 강의를 듣기 위해 두 달간 그곳에 머물렀다.

플램스티드가 더비셔에서 생활을 꾸리려고 하고 있을 때, 조너스

무어(Jonas Moore)가 그를 런던으로 초대하였다. 무어는 찰스 2세의 어린 시절 가정교사를 지냈고, 국방 목적으로 무기, 총기, 지도 등을 관리하는 법률 담당관으로 일하고 있었다. 그는 새로운 천문대 건설에 쓸 자금을 왕립학회에 요청 중이었다. 왕립학회는 당시 제기되던 경도를 결정하는 문제 해결을 위해 천문대 건설을 찰스 2세에게 간청했다. 하지만 그 사안은 국왕의 관심을 끌지 못했다.

찰스 2세는 프랑스 브르타뉴에서 온 포츠머스 공작부인 루이즈 드 케루아유(Louise de Kérouaille)를 애인으로 사귀고 있었다. 그녀는 자기가 아는 르 시외르 드 생 피에르(Le Sieur de St. Pierre)라는 아마추어 천문학자가 달의 위치로부터 경도를 알아낼 수 있다고 주장하였다. 국왕은 그녀의 말에 따라 1674년 12월 왕립위원회에 생 피에르의 주장을 조사하도록 명령했다.

플램스티드는 1675년 2월 2일 런던에 도착하여 무어와 함께 런던탑에서 지내고 있었다. 플램스티드의 천문 관측 경력을 알아본 위원회 덕분에 그는 왕을 알현하게 되었다. 왕립위원회의 요구로 플램스티드는 공식 증인으로 출석하였고, 생 피에르의 주장을 검증할 목적으로 실제 관측 자료들을 제시하였다. 그는 생 피에르의 형편없는 천문학 지식을 폭로했다. 그가 제공한 정보에 근거하여 작성된 왕립위원회 보고서의 결론은 크게 세 가지다.

첫째, 생 피에르의 주장은 일고의 가치도 없다.
둘째, 경도를 측정하는 간단한 방법은 아직 없다.

세상에서 가장 쉬운 과학 수업 별의 물리학

셋째, 왕은 더 정확한 천문 지도를 작성하고, 경도 측정에 요구되는 정밀도로 달의 움직임을 기록할 수 있도록 천문대 설립을 검토하여 그 적임자를 천문대장에 임명하여야 한다.

국왕의 대신이었던 무어는 수학과 천문학을 좋아했기 때문에 젊은 플램스티드를 알게 된 후 그를 후원하기 시작했다. 무어의 강력한 지원으로 1675년 3월 4일 28세의 플램스티드는 최초의 왕실 천문관, 당시 명칭으로 왕실 천문대장(The King's Astronomical Observator)에 임명되었다.

1675년 6월 22일 찰스 2세는 '완벽한 항해와 천문학을 위해 특정 위치의 경도를 찾는 목적'으로 왕립 천문대 건설을 약속했다. 그리고 천문대를 세울 장소로 그리니치의 왕실 영지와 520파운드의 건설비를 댄다. 찰스 2세는 천문학자들이 거주할 집과 다소 화려한 천문대를 설계하는 임무를 크리스토퍼 렌에게 위임했다. 1676년 2월 플램스티드는 왕립학회 회원 자격을 얻었고, 그해 7월 완공된 천문대의 초대 대장이 되어 1684년까지 그곳에서 지냈다.

영국 천문학자들은 오랫동안 그리니치 천문대를 위치 측정의 기준으로 삼아 왔다. 네 개의 자오선[4]이 그리니치 천문대를 기준으로 하며, 특히 경도의 기준인 본초자오선은 1851년에 정해졌고 1884년에는 국제회의를 통과했다. 이 자오선은 그리니치 공원에 황동(지금은

---

4) 북극점과 남극점을 최단 거리로 연결하는 지구 표면상의 세로선을 말한다.

스테인리스강)으로 표시되었다.

그리니치 천문대를 지나는 본초자오선(경도가 0인 위치)

정교수　다시 브래들리의 이야기로 돌아갈까? 1727년 당시 그리니치 천문대 대장인 브래들리는 광행차 현상을 최초로 발견했어.

물리군　광행차가 뭐예요?

정교수　별빛이 실제로 오는 방향과 관측되는 방향의 차이를 말해.

물리군　음…… 잘 모르겠어요.

정교수　우리는 정지해 있는 곳에서 별을 관측하는 것이 아니라 태양 주위를 도는 지구 위에서 별을 관측하네. 그래서 별이 실제 위치와 다른 곳에 있는 것처럼 보이는 현상을 브래들리가 처음 발견한 거지.

물리군　지구 공전의 증거이군요.

정교수　맞아. 광행차는 관측자가 움직이기 때문에 느껴지는 현상이야.

예를 들어 비가 오는 경우를 생각하자. 정지한 관측자는 빗방울이 똑바로 떨어지는 것을 관측한다고 하자.

만일 앞으로 뛰어가면 비가 비스듬히 내리는 것처럼 보이므로 우리는 우산을 앞으로 기울이게 된다.

이것은 움직이는 관측자가 비의 실제 속도를 보는 것이 아니라 상대속도를 보기 때문이다. 속도는 벡터이므로 벡터의 뺄셈에 의해 상대속도가 결정된다.

비의 실제 속도를 $\vec{u}$, 움직이는 관측자의 속도를 $\vec{v}$ 라고 하면 관측자가 느끼는 비의 상대속도는

$$\vec{u} - \vec{v}$$

이다.

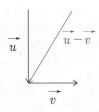

즉, 움직이는 관측자는 비가 비스듬히 내리는 것으로 보게 된다. 이 현상은 별빛의 방향을 관측할 때도 동일하게 적용된다. 다음 그림을 보자.

세상에서 가장 쉬운 과학 수업 별의 물리학

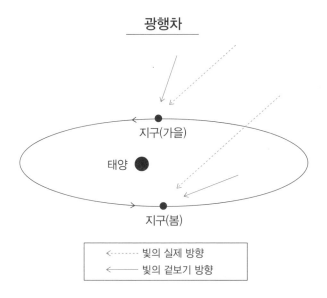

광행차

위 그림에서 점선은 별빛의 실제 진행 방향이고, 실선은 지구가 움직이기 때문에 지구의 관측자가 보는 별빛의 겉보기 방향이다. 즉, 지구의 관측자는 별의 실제 위치를 보지 못하고 겉보기 위치를 보게 된다. 이것을 광행차라고 부른다.

별까지의 거리 _베셀과 연주시차

물리군 별까지의 거리를 알 수 있나요?

정교수 물론이야. 별까지의 거리를 최초로 계산한 사람은 독일의 베셀이라네. 그의 삶과 연구에 대해 살펴볼까?

베셀(Friedrich Wilhelm Bessel, 1784~1846)

베셀은 1784년 독일의 민덴에서 태어났다. 그는 14세에 브레멘의 수출입 회사인 쿨렌캄프의 견습생이 되었다.

독학으로 수학과 천문학을 공부한 베셀은 1804년, 핼리 혜성의 정확한 궤도를 계산해 세상의 주목을 받았다. 1806년 그는 직장을 관두고 브레멘 근처에 있는 슈뢰터(Johann Hieronymus Schröter)의 개인 천문대 조수가 되었다. 그곳에서 그는 약 3,222개의 별에 대한 정확한 위치를 찾는 연구를 했다.

대학 교육을 받지 못했음에도 베셀은 1810년 1월 25세의 나이에 프로이센의 왕 프리드리히 빌헬름 3세(Friedrich Wilhelm Ⅲ)가 새로 설립한 쾨니히스베르크 천문대(Königsberg Observatory)의 책임자로 임명되어 죽을 때까지 그 직책을 유지했다.

1804년부터 1843년까지 베셀은 자신의 수학 및 천문학 연구 내용을 가우스와 편지로 주고받았다. 그는 가우스의 추천으로 1811년 3월 괴팅겐 대학교에서 명예박사 학위를 받았다.

세상에서 가장 쉬운 과학 수업 별의 물리학

쾨니히스베르크 천문대

베셀은 렌즈를 연마하고 관측 기술을 정교하게 다듬으면서 쾨니히
스베르크에서 28년을 보낸 후, 1838년에 마침내 연주시차를 측정하
는 데 성공했다. 6개월간의 고통스러운 관측과 분석 작업 끝에 그는
백조자리 61(61 Cygni)의 연주시차가 0.6272초, 즉 약 0.0001742도
라는 것을 알아냈다.

베셀이 측정한 연주시차는 매우 작은 값이었다. 그는 태양, 지구 그
리고 백조자리 61이 이루는 직각삼각형을 생각했다. 태양과 지구 사이
의 거리는 이미 알고 있고 측정에 의해 각도도 알게 되었으므로 삼각
함수를 이용하여 이 별까지의 거리를 계산할 수 있었다. 베셀의 측정에
의하면 백조자리 61까지의 거리는 10^{14}킬로미터(100조 킬로미터)나
되었다.

백조자리 61
(출처: Гоша0102/Wikimedia Commons)

물리군 베셀은 어떻게 별까지의 거리를 구한 거죠?

정교수 별까지의 거리를 재려면 먼저 연주시차를 알아야 해.

다음 그림을 보자.

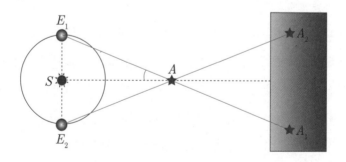

태양의 위치를 S, 별의 위치를 A라고 하자. 태양 주위를 공전하는 지구가 E_1의 위치에 있을 때 별은 A_1의 위치에 있는 것으로 보이고, 6

세상에서 가장 쉬운 과학 수업 별의 물리학

개월이 지나 지구가 E_2의 위치에 있을 때 별은 A_2의 위치에 있는 것으로 보인다. 이것은 지구가 공전하므로 별의 겉보기 위치가 달라지는 현상이다.

이때 $\angle E_1AS$를 연주시차라고 부른다. 연주(年周)라는 호칭이 붙는 것은 공전에 의해 생기는 시차인 까닭이다.

1543년 코페르니쿠스가 지동설을 발표한 이래, 천문학자들은 연주시차를 측정하려고 노력했다. 그것이 지구 공전에 대한 가장 확실하고도 직접적인 증거이기 때문이었다. 지동설 이후 3세기가 지나도록 수많은 사람이 도전했지만 연주시차는 측정되지 않았다. 천문학자 허셜도 평생을 바쳐 연주시차를 측정하려 했으나 실패했다. 하지만 베셀은 연주시차 측정에 성공했고 이를 통해 별까지의 거리를 구할 수 있었다.

다음 그림과 같이 연주시차가 θ인 경우를 보자.

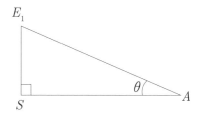

지구에서 별까지의 거리 r는 $\overline{AE_1}$ 의 길이이므로 삼각비의 정의에 따라

$$\sin\theta = \frac{\overline{E_1S}}{r}$$

이다. 그런데 $\overline{E_1S}$ 는 지구와 태양 사이의 거리이므로

$$\overline{E_1S} = 1.495978707 \times 10^{11}\,(\mathrm{m})$$

이고, 이것을 1AU(천문단위)로 쓴다. 따라서 거리를 천문단위로 계산하면

$$r = \frac{1}{\sin\theta}$$

이다.

실제로 관측된 연주시차는 매우 작은 각도였다. 아주 작은 각도를 나타낼 때는 초라는 각도 단위를 사용한다.

1도 = 60분

1분 = 60초

이렇게 각도의 단위인 도, 분, 초를 사용하여

1도 = 3600초

임을 알 수 있다. 이제 라디안이라는 각도 단위를 생각해 보자.

1라디안 = $\dfrac{180}{\pi}$ 도

세상에서 가장 쉬운 과학 수업 별의 물리학

θ를 라디안으로 나타낼 때 θ가 아주 작은 경우에는

$$\sin \theta \approx \theta$$

로 근사한다. 그러므로 별까지의 거리는

$$r = \frac{1}{\theta}$$

이 된다. 연주시차의 관측값은 초라는 단위를 사용하므로

$$\theta\text{라디안} = \phi\text{초}$$

라고 할 때, 비례식

$$1\text{라디안} : \frac{180}{\pi} \times 3600\text{초} = \theta\text{라디안} : \phi\text{초}$$

로부터

$$\theta = \frac{\pi\phi}{180 \times 3600}$$

이다. 즉, 연주시차가 ϕ초일 때 별까지의 거리는

$$r = \frac{206265}{\phi}(\text{AU})$$

임을 알 수 있다. 연주시차가 1초일 때 별까지의 거리는 206265 AU 이다. 천문학자들은 이 거리를 1파섹(pc)으로 정의했다.

1 pc = 206265 AU

천문학자들은 거리의 단위로 AU보다는 광년을 선호한다. 1광년은 빛의 속력으로 1년 동안 간 거리를 말하는데

1광년 = 약 63241 AU

이다. 그러므로

1 pc = 약 3.26광년

이 된다.

물리군 연주시차가 클수록 별까지의 거리가 가까워지는군요. 그럼 지구에서 제일 가까운 별은 뭐죠?

정교수 태양을 제외하고 지구에서 가장 가까운 별은 프록시마 켄타우리일세. 켄타우루스자리에 있는 별이야. 프록시마(proxima)는 '인접해 있다, 가깝다'라는 뜻이지. 붉은색을 띠는 프록시마 켄타우리는 너무 어두워서 맨눈으로는 볼 수 없어. 이 별의 연주시차는 0.768초라네. 그러니까 프록시마 켄타우리까지의 거리는 4.22광년 정도야. 두 번째로 가까우면서 맨눈으로 볼 수 있는 별은 알파 켄타우리야. 이 별의 연주시차는 0.747초이므로 알파 켄타우리는 지구로부터 4.35광년 거리에 있지.

세상에서 가장 쉬운 과학 수업 별의 물리학

두 밝은 별 중 왼쪽은 알파 켄타우리, 오른쪽은 베타 켄타우리이고 원으로 표시한 희미한 별이 프록시마 켄타우리이다.(출처: Skatebiker at English Wikipedia)

별의 절대등급_별의 실제 밝기 비교하기

정교수 　별까지의 거리를 잴 수 있게 되면서 별의 절대등급에 대한 연구가 시작되었어.

물리군 　절대등급이 뭔가요?

정교수 　첫 번째 만남에서 언급했던 히파르코스가 별을 여섯 등급으로 나눈 이야기를 다시 떠올려 볼까?

히파르코스가 나눈 별의 등급은 우리가 보는 별의 겉보기등급이다. 별까지의 거리는 모두 같지 않으므로 실제로 아주 밝은 별이라도 멀리 떨어져 있으면 어두운 별로 보인다.

별의 겉보기등급은 밝기에 대한 로그를 이용해서 정의한다. 지구로부터 거리가 r인 별의 밝기가 L일 때, 이 별의 겉보기등급 m은 로그에 의해 다음과 같이 정의한다.

$$m = -2.5 \log_{10} \frac{I_r}{I_0}$$

(2-7-1)

또는

$$I_r = 10^{-0.4m} I_0$$

(2-7-2)

여기서 I_0은 $2.48 \times 10^{-8} (\text{W/m}^2)$의 밝기이고 W는 일률의 단위인 와트이다.

이제 히파르코스가 분류한 1등성과 6등성의 밝기의 비를 알아보자. 식 (2-7-2)로부터

$$I_{1등성} = 10^{-0.4} I_0$$

$$I_{6등성} = 10^{-0.4 \times 6} I_0$$

이므로

$$\frac{I_{1등성}}{I_{6등성}} = 10^2 = 100$$

이 되어, 1등성의 밝기는 6등성의 밝기의 100배이다.

겉보기등급으로는 별의 실제 밝기를 알 수 없다. 이런 필요에서 절

세상에서 가장 쉬운 과학 수업 별의 물리학

대등급이 도입되었다. 별이 얼마나 밝은지를 정확하게 판단하려면 별들을 같은 거리에 놓고 비교해야 한다.

천문학자들은 별을 지구로부터 10파섹(32.6광년)의 거리에 놓는 것으로 가정할 때 지구에서 보이는 밝기를 별의 절대등급이라고 부른다. 절대등급이 작을수록 실제로 밝은 별을 나타낸다.

겉보기등급의 경우는 어두운 별이라도 지구 가까이에 있으면 등급이 높아진다. 하지만 절대등급은 별을 같은 거리에 놓고 밝기를 비교하므로 별까지의 거리는 영향을 미치지 않는다. 예를 들어 태양의 겉보기등급은 무려 −26.74등급으로 엄청 높지만, 절대등급은 고작 4.83등급에 불과하다.

물리군   별의 등급이 음수가 나올 수도 있군요.
정교수   그렇지. 0등급보다 밝은 별은 음수의 등급을 가진다네.

지구로부터 거리가 r인 별의 밝기가 I_r일 때, 이 별이 10 pc의 거리에 있을 때의 별의 밝기를 I_{10}이라고 하면 이 별의 절대등급 M은 로그에 의해 다음과 같이 정의한다.

$$M = -2.5 \log_{10} \frac{I_{10}}{I_0}$$

(2-7-3)

별의 밝기는 거리의 제곱에 반비례한다.

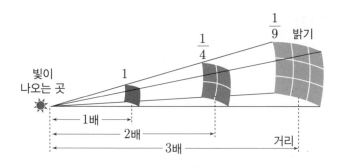

$$별의\ 밝기 \propto \frac{1}{(별까지의\ 거리)^2}$$

거리에 따른 별의 밝기 변화

지구로부터 거리가 r인 별의 밝기와 거리가 $10\,\mathrm{pc}$인 별의 밝기의 비는

$$I_r : I_{10} = \frac{1}{r^2} : \frac{1}{10^2}$$

이므로

$$\frac{I_r}{I_{10}} = \frac{100}{r^2} \tag{2-7-4}$$

이다. 따라서 별의 겉보기등급과 절대등급의 차이는

$$m - M = -2.5 \log_{10} \frac{I_r}{I_{10}} \tag{2-7-5}$$

세상에서 가장 쉬운 과학 수업 별의 물리학

가 된다. 식 (2-7-4)를 식 (2-7-5)에 넣으면

$$m - M = -5 + 5 \log_{10} r \qquad\qquad (2\text{-}7\text{-}6)$$

임을 알 수 있다.

세 번째 만남

•

폴리트로픽 과정과
별의 구조 방정식

폴리트로픽 과정_별 속의 기체의 열역학

정교수 지금부터는 별의 구조 방정식에 대해 설명하겠네. 그에 앞서 별 속의 기체들의 압력과 밀도 사이의 관계를 확인할 필요가 있어. 그러려면 별 속의 기체의 열역학을 조금 알아두어야 하지.

물리군 기체의 열역학이 뭔가요?

정교수 기체는 열을 얻으면 온도가 올라가고 팽창하네. 이런 현상을 다루는 것이 기체의 열역학이라고 생각하면 돼. 이제 우리는 이상기체를 고려할 거야.

물리군 이상기체는 뭐죠?

정교수 기체 분자들 사이의 퍼텐셜 에너지를 0으로 가정한 이상적인 기체를 말해. 즉, 이상기체의 역학적 에너지는 운동 에너지가 되지.

열역학에는 두 개의 기본 법칙이 나온다. 바로 열역학 제1법칙과 열역학 제2법칙이다. 이 중 열역학 제1법칙을 살펴보자.

[열역학 제1법칙]
계에 공급된 열[5]의 변화량은 계의 내부 에너지의 변화와 계가 한 일의 합이다.

---

5) 계에서 열이 빠져나간 경우는 계에 음수의 열이 공급되었다고 생각하면 된다.

우리는 이상기체를 다루므로 내부 에너지는 기체 분자들의 운동 에너지의 합이다. 즉, 내부 에너지는 계의 에너지의 총합을 말한다.

열을 Q, 내부 에너지를 U, 일을 W라고 하면 열역학 제1법칙은

$$\Delta Q = \Delta U + \Delta W \qquad\qquad (3\text{-}1\text{-}1)$$

로 쓸 수 있다. 앞으로는 계에 열이 흘러 들어간 경우를 생각하자.

물리군 여러 번 배웠는데 계라는 용어가 낯설어요.

정교수 계는 우리가 관심 있어 하는 대상을 뜻해. 별의 물리학에서는 별이 계가 되는 거지.

물리군 조금 이해가 되네요.

정교수 계가 일을 하지 않으면 공급된 열이 모두 계의 내부 에너지 증가에 사용돼. 하지만 계가 일을 하면 공급된 열 중에서 일로 사용된 것을 제외한 나머지 에너지가 계의 내부 에너지를 증가시킨다는 것이 열역학 제1법칙이야.

물리군 계는 어떻게 일을 하죠?

정교수 예를 들어 풍선을 뜨겁게 하는 경우를 살펴볼까?

풍선이라는 계로 열이 공급된 경우를 생각하자. 이때 풍선은 팽창한다. 이렇게 풍선이 팽창하려면 풍선 속의 기체 분자들이 풍선 벽을 밀어야 한다. 즉, 풍선 벽에 힘을 작용해야 한다. 여기서 힘이 한 일은 다음과 같다.

(힘이 한 일) = (힘) × (풍선을 밀어낸 거리)

= (압력) × (단면적) × (풍선을 밀어낸 거리)

= (압력) × (팽창한 부피)

압력을 p, 부피를 V라고 하면

$$\Delta W = p \Delta V \tag{3-1-2}$$

이므로 열역학 제1법칙으로부터

세상에서 가장 쉬운 과학 수업 별의 물리학

$$\Delta Q = \Delta U + p\Delta V \qquad (3\text{-}1\text{-}3)$$

가 된다.

이제 열용량과 비열을 알아보자. 어떤 물질의 온도를 1도 올리기 위한 열에너지의 양을 열용량이라고 하며 C로 쓴다. 계에 열이 ΔQ만큼 공급되어 온도가 ΔT만큼 변했을 때 열용량은

$$C = \frac{\Delta Q}{\Delta T}$$

이고, 이를 바꾸어 쓰면

$$\Delta Q = C\Delta T \qquad (3\text{-}1\text{-}4)$$

이다. 열용량은 물질의 온도를 1도 올리는 데 필요한 열의 양이므로 같은 물질이라도 양에 따라 차이가 난다. 예를 들어 큰 욕조의 물이 컵의 물보다 열용량이 크다.

그래서 공정한 비교를 위해 비열을 정의한다. 비열은 단위질량[6]당 열용량으로 c라고 쓴다. 단위질량의 기체를 고려하면 식 (3-1-4)는

$$\Delta Q = c\Delta T \qquad (3\text{-}1\text{-}5)$$

가 된다.

비열에는 두 종류가 있다. 부피가 일정할 때의 비열을 일정부피비

---

6) MKS 단위계에서 단위질량은 1kg이다.

열이라 하고 c_V로 쓰며, 압력이 일정할 때의 비열을 일정압력비열이라 하고 c_p로 쓴다.

이제 단위질량의 이상기체를 생각하자. 1834년 프랑스의 과학자 클라페롱(Benoît Paul Émile Clapeyron, 1799~1864)은 기체의 종류와 상관없이 기체 n몰의 경우 온도(T), 부피(V), 압력(p)은 다음 관계를 만족하는 것을 알아냈다.

$$pV = nRT \tag{3-1-6}$$

이 식을 이상기체 방정식이라고 부른다. 여기서 R는 기체상수라고 하며 기체의 종류와 관계없이

$$R = 8.31446261815324 (\mathrm{JK^{-1}mol^{-1}})$$

이다.

물리군 몰수가 뭐죠?

정교수 기체 1몰 속에는 아보가드로수만큼의 기체 분자가 있어. 아보가드로수는 N_A로 쓰는데

$$N_A = 6.02214076 \times 10^{23}$$

이지. 그러니까 기체 n몰 속에는 기체 분자가 아보가드로수의 n배 만큼 있다네.

세상에서 가장 쉬운 과학 수업 별의 물리학

물리군 그렇군요.

정교수 단원자 이상기체[7]의 경우, 내부 에너지 변화는 온도 T의 변화에 비례해. 이것은 영국의 물리학자 맥스웰에 의해 알려졌지. 기체 n몰의 경우 내부 에너지는

$$U = \frac{3}{2}nRT \qquad\qquad (3\text{-}1\text{-}7)$$

라네. 그러니까 온도가 ΔT만큼 변할 때 내부 에너지의 변화는

$$\Delta U = \frac{3}{2}nR\Delta T \qquad\qquad (3\text{-}1\text{-}8)$$

가 되지.

먼저 일정부피비열을 구해 보자. 부피가 일정한 경우 $\Delta V = 0$이므로 열역학 제1법칙으로부터

$$\Delta Q = \Delta U = \frac{3}{2}nR\Delta T = c_V\,\Delta T \qquad\qquad (3\text{-}1\text{-}9)$$

이다. 따라서 일정부피비열은

$$c_V = \frac{3}{2}nR \qquad\qquad (3\text{-}1\text{-}10)$$

---

7) 헬륨, 네온처럼 원자 하나가 분자를 이루는 기체를 말한다. 원자 두 개가 모여 분자를 만드는 수소나 산소는 이원자 기체라고 부른다.

이다. 이제 일정압력비열을 구하자. 이상기체 방정식으로부터

$$\Delta(pV) = nR\Delta T$$

이고, p가 일정하므로

$$p\Delta V = nR\Delta T \tag{3-1-11}$$

가 된다. 식 (3-1-8), (3-1-11)을 식 (3-1-3)에 넣으면

$$\Delta Q = \Delta U + p\Delta V$$

$$= \frac{3}{2}nR\Delta T + nR\Delta T$$

$$= \frac{5}{2}nR\Delta T$$

가 되어,

$$c_p = \frac{5}{2}nR \tag{3-1-12}$$

임을 알 수 있다.

이때 일정부피비열에 대한 일정압력비열의 비를

$$\gamma = \frac{c_p}{c_V} \tag{3-1-13}$$

로 정의하면 단원자 이상기체의 경우

세상에서 가장 쉬운 과학 수업 별의 물리학

$$\gamma = \frac{5}{3}$$

가 된다. 일반적으로 일정부피비열과 일정압력비열의 관계는

$$c_p = c_V + nR \tag{3-1-14}$$

이다.

이번에는 단열과정을 생각하자. 단열과정이란 계에 공급되는 열의 변화량이 0인 경우를 말한다. 즉, $\Delta Q = 0$이므로 열역학 제1법칙에 따라

$$\Delta U = -p\Delta V \tag{3-1-15}$$

이다. 아주 작은 부피 변화(dV)를 고려하면 위 식은

$$dU = -pdV \tag{3-1-16}$$

가 되므로 단열과정의 경우 압력은

$$p = -\frac{\partial U}{\partial V} \tag{3-1-17}$$

로 나타낼 수 있다.

한편 이상기체 방정식으로부터

$$\Delta(pV) = nR\Delta T$$

이므로

$$p \varDelta V + V \varDelta p = nR \varDelta T \qquad (3-1-18)$$

가 된다.

또한 일정부피비열의 정의로부터

$$\varDelta T = \frac{\varDelta U}{c_V} = -\frac{p \varDelta V}{c_V}$$

이다. 그러므로 식 (3-1-18)에서

$$p \varDelta V + V \varDelta p = -\frac{nR}{c_V} p \varDelta V$$

이고, 정리하면

$$\gamma p \varDelta V + V \varDelta p = 0 \qquad (3-1-19)$$

이 된다. 여기서

$$\gamma = 1 + \frac{nR}{c_V} \qquad (3-1-20)$$

이다.

이제 p와 V의 관계가 다양한 경우를 생각하자. 그중에서 p가 V^α에 비례하는 모델을 폴리트로픽(polytropic) 과정이라고 부른다. 이때 비례상수를 A로 놓으면

$$p = AV^\alpha \qquad (3-1-21)$$

이다. 예를 들어 $\alpha = 0$이면 일정압력과정을, $\alpha = -1$이면 일정온도과정을 나타낸다.

폴리트로픽 과정의 경우

$$\Delta p = A\alpha V^{\alpha-1} \Delta V \qquad (3-1-22)$$

이므로 식 (3-1-19)에 대입하면

$$\gamma A V^{\alpha} \Delta V + V A\alpha V^{\alpha-1} \Delta V = 0$$

또는

$$(\gamma + \alpha)A V^{\alpha} \Delta V = 0 \qquad (3-1-23)$$

이 된다. 즉,

$$\alpha = -\gamma \qquad (3-1-24)$$

이다. 따라서 폴리트로픽 단열팽창의 경우 압력과 부피의 관계는

$$pV^{\gamma} = A = (일정) \qquad (3-1-25)$$

하다.

이제 부피당 기체 분자의 개수를 기체의 밀도 ρ로 정의하면

$$\rho = \frac{N}{V}$$

이다. N은 일정하므로 식 (3-1-25)는

$$p = AN^{-\gamma}\rho^{\gamma} \qquad\qquad (3\text{-}1\text{-}26)$$

으로 쓸 수 있다. 여기서

$$K = AN^{-\gamma} \qquad\qquad (3\text{-}1\text{-}27)$$

으로 놓으면

$$p = K\rho^{\gamma} \qquad\qquad (3\text{-}1\text{-}28)$$

이 된다. 이것이 폴리트로픽 단열과정(줄여서 폴리트로픽 과정)의 기본 관계식이다.

예를 들어 단원자 이상기체의 폴리트로픽 과정은

$$pV^{\frac{5}{3}} = A = (\text{일정})$$

또는

$$p = K\rho^{\frac{5}{3}}$$

에 의해 묘사된다.

레인과 엠덴의 폴리트로픽 별 모형_별이 평형을 이루기 위한 방정식

정교수   이제 우리는 별의 물리학을 최초로 연구한 미국의 천체물리학자 레인(Jonathan Homer Lane, 1819~1880)을 알아볼 거야.

레인은 미국 뉴욕주의 제너시오에서 태어났다. 예일 대학에 다니던 시절 그는 기상학자 데니슨 옴스테드(Denison Olmsted) 밑에서 학업을 수행했다. 또한 에스피(James P. Espy)의 폭풍에 대한 열역학 모델을 공부했다.

1841년 에스피의 책 《폭풍의 철학》
에 수록된 그림

1846년에 대학을 졸업한 후 레인은 미국 특허청에서 근무했으며, 1851년에 수석 심사관이 되어 1857년까지 그 일을 계속했다. 1860년부터 1866년까지 그는 펜실베이니아주 프랭클린에서 대장장이인 형과 함께 살았다. 그 기간 동안 레인은 영하 209도의 저온을 만드는 장치 개발에 적극적으로 참여했다.

레인은 특히 천문학에 관심이 많았다. 그는 별 속의 기체의 압력, 온도, 밀도 사이의 열역학적 관계를 연구해, 1870년에 평형을 이루는 별 구조에 관한 논문을 발표했다.

물리군 최초로 별의 물리학을 연구한 학자인데 사진이 없네요.

정교수 레인의 연구가 사람들에게 크게 알려지지 않아서 그렇다네. 그의 연구는 별에 대한 본격적인 연구가 시작되는 1900년대보다 30년이나 앞선 시기에 발표되었거든.

레인의 연구를 확장한 별의 모형 연구는 스위스의 엠덴(Jacob Robert Emden, 1862~1940)에 의해 이루어진다.

엠덴은 스위스 장크트갈렌에서 태어났다. 그는 하이델베르크와 베를린에서 수학과 물리학을 공부했고, 1885년에 물리학 학사 학위를, 1887년 스트라스부르 대학에서 물리학 박사 학위를 받았다. 그의 박사 학위 논문은 염 용액의 증기압에 관한 내용이었다.

1889년 엠덴은 뮌헨 공과대학 물리학 교수로 임명되어 기상학을 연구했다. 1907년에는 《천문학과 기상학에 대한 열역학 모델》이라는 책을 출간했다. 이 책에서 그는 별 속의 기체들을 폴리트로픽 기체로 간주하고 별이 평형을 이루기 위한 방정식을 발표했다. 또한 레인의 논문을 소개했는데 이로 인해 이 방정식은 레인-엠덴 방정식으로 불린다.

세상에서 가장 쉬운 과학 수업 별의 물리학

물리군   별이 평형을 이룬다는 게 무슨 말인지 잘 모르겠어요.

정교수   별 속의 기체에 작용하는 두 힘이 크기가 같고 방향이 반대
면 두 힘이 평형을 이루는 걸세.

　레인-엠덴 방정식을 자세히 살펴보자. 별을 반지름이 R이고 질량
이 M인 공 모양이라고 하자. 별의 중심에서 거리 r만큼 떨어진 곳의
중력가속도의 크기를 $g(r)$라고 하면, 식 (2-2-5)로부터

$$g(r) = \frac{GM(r)}{r^2} \tag{3-2-1}$$

가 된다. 여기서 $M(r)$는 반지름이 r인 구 속의 질량을 의미한다.

　별 속의 밀도가 일정하지 않고 중심으로부터의 거리에 따라 달라
지는 경우를 생각하자. 별의 중심에서 거리 r만큼 떨어진 곳의 밀도
를 $\rho(r)$라고 하면

$$M(r) = \int_{\text{반지름 } r \text{인 구}} \rho(r')\, dv' \tag{3-2-2}$$

이 된다. dv'은 구에서의 아주 작은 부피를 나타내는데, 이것을 부피
요소라고 한다.

물리군   구에서의 부피 요소는 어떻게 구하나요?

정교수   구와 잘 어울리는 좌표계를 사용해야 해. 이를 구좌표계라고
부르지.

다음 그림을 보자.

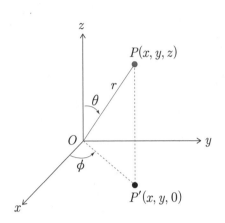

구좌표계는 원점 O에서 점 P까지의 거리 r, z축과 $\overline{OP}$ 의 사잇각 θ, 점 P의 xy평면으로의 수선의 발을 $P'(x, y, 0)$이라고 할 때 $\overline{OP'}$ 과 x축의 양의 방향이 이루는 각 ϕ로 구성된다. 즉, 점 P를 구좌표로 나타내면

$P(r, \theta, \phi)$

가 된다. 여기서 θ를 편각, ϕ를 방위각이라고 한다. 이때 편각의 범위는

$0 \leq \theta \leq \pi$

이다. 예를 들어 북극점의 편각은 $\theta = 0$이고 남극점의 편각은 $\theta = \pi$이다. 방위각의 범위는 다음과 같다.

$$0 \le \phi \le 2\pi$$

구좌표계로 나타낸 점 (r, θ, ϕ)에서 구와 잘 어울리는 아주 작은 조각을 그리면 다음 그림과 같다.

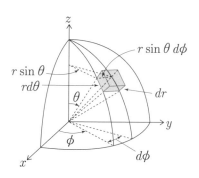

우리가 dr, $d\theta$, $d\phi$를 거의 0에 가깝게 작게 택했으므로 위 그림에서 보이는 조각은 직육면체와 같다.

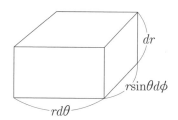

이 입체의 부피가 바로 구에서의 부피 요소 dv이다. 즉, 부피 요소는

$$dv = (dr) \times (rd\theta) \times (r\sin\theta d\phi) = r^2 \sin\theta dr d\theta d\phi \qquad (3\text{-}2\text{-}3)$$

이다.

물리군 반지름이 R인 구 속의 부피 요소를 모두 더하면 구의 부피가 나오겠군요!

정교수 맞아. 연속적인 양을 더하는 것을 적분이라고 부르지. 이 경우는 적분이 3개 필요해. r에 대한 적분, θ에 대한 적분, ϕ에 대한 적분, 이렇게 말일세. r에 대한 적분에서는 r가 0부터 R까지, θ에 대한 적분에서는 θ가 0부터 π까지, ϕ에 대한 적분에서는 ϕ가 0부터 2π까지 변하지. 그러니까 반지름이 R인 구의 부피를 V라고 하면

$$
\begin{aligned}
V &= \int_{r=0}^{R}\int_{\theta=0}^{\pi}\int_{\phi=0}^{2\pi} dv \\
&= \int_{r=0}^{R}\int_{\theta=0}^{\pi}\int_{\phi=0}^{2\pi} r^2 \sin\theta dr d\theta d\phi \\
&= \left(\int_{r=0}^{R} r^2 dr\right)\left(\int_{\theta=0}^{\pi} \sin\theta d\theta\right)\left(\int_{\phi=0}^{2\pi} d\phi\right) \\
&= \frac{R^3}{3} \times 2 \times 2\pi \\
&= \frac{4}{3}\pi R^3
\end{aligned}
$$

이 된다네.

세상에서 가장 쉬운 과학 수업 별의 물리학

물리군 구의 부피 공식이 나오네요.

정교수 당연하지. 이제 식 (3-2-2)는

$$M(r) = \int_{\text{반지름 } r \text{인 구}} \rho(r')\, dv'$$

$$= \int_{\text{반지름 } r \text{인 구}} \rho(r')\, r'^2 \sin\theta'\, dr'\, d\theta'\, d\phi'$$

$$= \left(\int_{r'=0}^{r} \rho(r')\, r'^2\, dr' \right) \left(\int_{\theta'=0}^{\pi} \sin\theta'\, d\theta' \right) \left(\int_{\phi'=0}^{2\pi} d\phi' \right)$$

$$= 4\pi \left(\int_{r'=0}^{r} \rho(r')\, r'^2\, dr' \right) \tag{3-2-4}$$

으로 쓸 수 있어. 식 (3-2-4)를 r로 미분하면 다음과 같아.

$$\frac{dM(r)}{dr} = 4\pi r^2 \rho(r) \tag{3-2-5}$$

이제 별 속에서 질량이 $\rho(r)dv$인 부분을 생각하자.

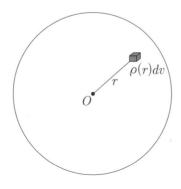

이 부분이 받는 중력의 크기는

$$\rho(r)\,dv \times g(r) = \frac{G\rho(r)M(r)}{r^2}dv \qquad (3\text{-}2\text{-}6)$$

이고, 방향은 별의 중심 방향이다.

레인과 엠덴은 별이 공 모양을 유지하려면 별의 중심 방향의 힘과 평형을 이루는 별의 바깥 방향으로 향하는 힘이 존재해야 한다고 생각했다. 그들은 이 힘을 기체의 압력에 의한 힘으로 보았다.

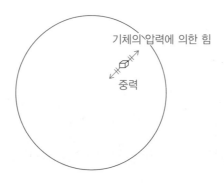

구에서의 아주 작은 입체 조각의 아랫면에 작용하는 압력을 $p(r)$, 윗면에 작용하는 압력을 $p(r+dr)$라고 하자.

　세상에서 가장 쉬운 과학 수업 별의 물리학

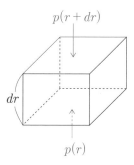

$p(r+dr)$

dr

$p(r)$

기체의 압력에 의한 힘이 별의 중심에서 바깥을 향하려면

$$p(r) > p(r+dr)$$

가 되어야 한다.

이제

$$p(r+dr) = p(r) + dp(r) \qquad\qquad (3\text{-}2\text{-}7)$$

라고 하면 $dp(r)$는 음수이다. 따라서 이 힘의 크기는

$$- dp(r) \times A$$

이다. 여기서 윗면과 아랫면의 넓이는 거의 같으므로 그 넓이를 A로 놓았다.

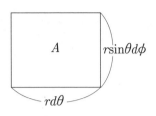

이 넓이는

$$A = r^2 \sin \theta d\theta d\phi \qquad (3\text{-}2\text{-}8)$$

이다. 기체의 압력에 의한 힘과 중력이 평형을 이루어야 하므로

$$-dp(r) \times A = \frac{G\rho(r)M(r)}{r^2}dv \qquad (3\text{-}2\text{-}9)$$

이고,

$$dv = dr \times A \qquad (3\text{-}2\text{-}10)$$

이므로

$$-dp(r) = \frac{G\rho(r)M(r)}{r^2}dr$$

또는

$$\frac{dp}{dr} = -\frac{GM(r)\rho(r)}{r^2} \qquad (3\text{-}2\text{-}11)$$

세상에서 가장 쉬운 과학 수업 별의 물리학

가 된다. 이 식은

$$\frac{r^2}{\rho}\frac{dp}{dr} = -GM(r)$$

(3-2-12)

라고 쓸 수 있다. 이를 미분하면

$$\frac{d}{dr}\left(\frac{r^2}{\rho}\frac{dp}{dr}\right) = -G\frac{dM(r)}{dr}$$

(3-2-13)

이다. 여기에 식 (3-2-5)를 넣으면

$$\frac{d}{dr}\left(\frac{r^2}{\rho}\frac{dp}{dr}\right) = -4\pi Gr^2\rho$$

또는

$$\frac{1}{r^2}\frac{d}{dr}\left(\frac{r^2}{\rho}\frac{dp}{dr}\right) = -4\pi G\rho$$

(3-2-14)

가 된다.

레인과 엠덴은 별 속의 기체를 폴리트로픽 기체로 간주했다. 그러므로

$$p = K\rho^\gamma$$

(3-2-15)

으로 놓을 수 있다. 여기서 우리는

$$\gamma = \frac{n+1}{n} \qquad (3\text{-}2\text{-}16)$$

로 두고, n을 폴리트로픽 지수라고 부른다. 이때

$$\frac{n+1}{n} \frac{K}{r^2} \frac{d}{dr}\left(r^2 \rho^{\frac{1-n}{n}} \frac{d\rho}{dr}\right) = -4\pi G\rho \qquad (3\text{-}2\text{-}17)$$

또는

$$\frac{\gamma K}{r^2} \frac{d}{dr}\left(r^2 \rho^{\gamma-2} \frac{d\rho}{dr}\right) = -4\pi G\rho \qquad (3\text{-}2\text{-}18)$$

가 되는데 이 방정식은 n에 따라 달라진다.

물리군　굉장히 복잡한 방정식이네요.

정교수　이걸 조금 더 간단한 모양으로 만들 수 있어.

식 (3-2-17)에서

$$\rho = \rho_c D^k \qquad (3\text{-}2\text{-}19)$$

으로 놓자. 그러면 식 (3-2-18)은

$$\frac{\gamma k K}{r^2} \frac{d}{dr}\left(r^2 \rho_c^{\gamma-1} D^{k(\gamma-1)-1} \frac{dD}{dr}\right) = -4\pi G\rho_c D^k \qquad (3\text{-}2\text{-}20)$$

으로 쓸 수 있다. 여기서

$$D^{k(\gamma-1)-1} = 1$$

이 되도록 k를 선택하자. 이것은

$$k(\gamma - 1) = 1$$

을 의미하므로

$$k = \frac{1}{\gamma - 1} = n \tag{3-2-21}$$

이다. 따라서 식 (3-2-20)은

$$\frac{\gamma n K \rho_c^{\gamma-2}}{r^2} \frac{d}{dr}\left(r^2 \frac{dD}{dr}\right) = -4\pi G D^n \tag{3-2-22}$$

이 된다. 이 식을 좀 더 예쁘게 만들기 위해

$$r = \lambda_n \xi \tag{3-2-23}$$

라고 두자. 이때

$$\lambda_n = \left[(n+1)\left(\frac{K\rho_c^{\frac{1-n}{n}}}{4\pi G}\right)\right]^{\frac{1}{2}} \tag{3-2-24}$$

이다. 그러므로 식 (3-2-22)는

$$\frac{1}{\xi^2}\frac{d}{d\xi}\left(\xi^2\frac{dD_n(\xi)}{d\xi}\right)=-D_n(\xi)^n \qquad (3\text{-}2\text{-}25)$$

이 된다. D_n은 n에 대응하는 D를 나타내며, 밀도와 D_n의 관계는 다음과 같다.

$$\rho(r)=\rho_c[D_n(\xi)]^n, \quad \xi=\frac{r}{\lambda_n} \qquad (3\text{-}2\text{-}26)$$

여기서 식 (3-2-25)를 레인-엠덴 방정식이라고 부른다.

별의 중심은 $r=0$에 대응한다. 이제 $D_n(0)=1$로 놓으면 ρ_c는 별의 중심부의 밀도가 된다.

별의 표면은 $r=R$에 대응하는데, 이곳에서의 별의 밀도를 0이라고 하자. $r=R$는 $\xi=\dfrac{R}{\lambda_n}$를 의미하므로

$$\rho(R)=\rho_c\left[D_n\left(\frac{R}{\lambda_n}\right)\right]^n=0$$

이다. 따라서

$$D_n\left(\frac{R}{\lambda_n}\right)=0 \qquad (3\text{-}2\text{-}27)$$

이 된다. $\xi_{1,n}=\dfrac{R}{\lambda_n}$ 라고 하면

$$D_n(\xi_{1,n})=0 \qquad (3\text{-}2\text{-}28)$$

이다.

한편 별의 중심부의 밀도는 변하지 않으므로

$$\frac{dD_n}{d\xi}\Big|_{\xi=0} = 0 \qquad (3\text{-}2\text{-}29)$$

이 성립한다.

물리군 레인-엠덴 방정식은 어떻게 푸나요?

정교수 레인-엠덴 방정식은 특별한 n에 대해서만 풀 수 있다네. $n = 0, 1, 5$에 대해서만 정확한 해가 구해지지.

먼저 $n = 0$인 경우를 보자. 이때 레인-엠덴 방정식은

$$\frac{1}{\xi^2}\frac{d}{d\xi}\left(\xi^2 \frac{dD_0}{d\xi}\right) = -1 \qquad (3\text{-}2\text{-}30)$$

또는

$$\frac{d}{d\xi}\left(\xi^2 \frac{dD_0}{d\xi}\right) = -\xi^2 \qquad (3\text{-}2\text{-}31)$$

이므로

$$\xi^2 \frac{dD_0}{d\xi} = -\frac{\xi^3}{3} + c_1 \qquad (3\text{-}2\text{-}32)$$

또는

$$\frac{dD_0}{d\xi} = -\frac{\xi}{3} + \frac{c_1}{\xi^2} \tag{3-2-33}$$

이 된다. $\xi \to 0$일 때 $\frac{dD_0}{d\xi} \to 0$이므로

$c_1 = 0$

이다. 즉,

$$\frac{dD_0}{d\xi} = -\frac{\xi}{3} \tag{3-2-34}$$

가 되어,

$$D_0 = -\frac{\xi^2}{6} + c_2 \tag{3-2-35}$$

로 쓸 수 있다. $D_0(0) = 1$이므로

$c_2 = 1$

이 되어,

$$D_0(\xi) = 1 - \frac{\xi^2}{6} \tag{3-2-36}$$

이 나온다. 이제 ξ_1을 구하자. 이것은

$$D_0(\xi_1) = 0$$

을 만족하므로

$$\xi_1 = \sqrt{6}$$

임을 알 수 있다. D_0의 그래프는 다음과 같다.

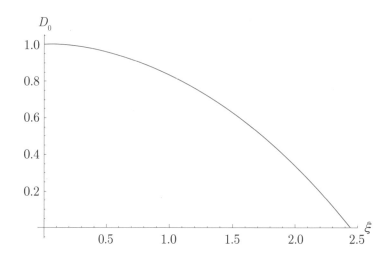

물리군 $n = 1$인 경우 레인-엠덴 방정식은

$$\frac{1}{\xi^2}\frac{d}{d\xi}\left(\xi^2\frac{dD_1}{d\xi}\right) = -D_1 \tag{3-2-37}$$

이 되는군요. 이것은 어떻게 풀죠?

정교수   이 방정식의 해를

$$D_1 = \xi^A u \qquad (3\text{-}2\text{-}38)$$

로 놓고, 식 (3-2-37)에 대입하여 정리하면

$$A(A+1)\xi^A u + 2(A+1)\xi^{A+1}u' + \xi^{A+2}u'' = -\xi^{A+2}u$$

가 돼. 이 식에서 $A = -1$을 선택하면

$$u'' = -u \qquad (3\text{-}2\text{-}39)$$

를 구할 수 있지.

물리군   식 (3-2-39)를 만족하는 u는 $\sin\xi$ 또는 $\cos\xi$겠네요.

정교수 $\cos\xi$는 안 돼.

물리군   그건 왜죠?

정교수 u가 $\cos\xi$이면

$$D_1 = \frac{\cos\xi}{\xi} \qquad (3\text{-}2\text{-}40)$$

인데 이것은 ξ가 0으로 갈 때 무한대가 되거든.

물리군 ξ가 0으로 갈 때 1이 되어야 하기 때문이군요.

정교수   그렇지. 그러니까 방정식 (3-2-37)의 해는

$$D_1(\xi) = \frac{\sin\xi}{\xi}$$

세상에서 가장 쉬운 과학 수업 별의 물리학

라네. 이것은 ξ가 0으로 갈 때 1이 돼. 그리고 $D_1(\xi_1) = 0$으로부터

$$\xi_1 = \pi$$

이지. D_1의 그래프는 다음과 같아.

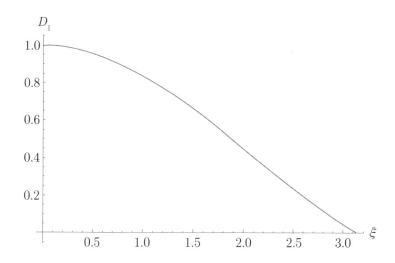

물리군　$n = 5$인 경우 레인-엠덴 방정식은

$$\frac{1}{\xi^2} \frac{d}{d\xi}\left(\xi^2 \frac{dD_5}{d\xi} \right) = -D_5^5 \tag{3-2-41}$$

이 되네요. 이것은 어떻게 푸나요?

정교수　이 방정식의 해를 일반적으로 구하는 방법은 없어. 여러 가지 시도를 해야 한다네.

물리학자들은 여러 번 시도한 끝에

$$D_5 = (1 + a\xi^2)^k \qquad\qquad (3-2-42)$$

의 모양을 찾아냈다. 이때

$$\frac{1}{\xi^2}\frac{d}{d\xi}\left(\xi^2\frac{dD_5}{d\xi}\right) = 2ak\left[3 + a\left(1 + 2k\right)\xi^2\right]\left(1 + a\xi^2\right)^{-2+k} \quad (3-2-43)$$

이 되고,

$$-D_5^5 = -\left(1 + a\xi^2\right)^{5k} \qquad\qquad (3-2-44)$$

이다. 식 (3-2-42)와 (3-2-43)에서

$$-2 + k = 5k$$

를 선택하면

$$k = -\frac{1}{2}$$

이므로 식 (3-2-41)로부터

$$a = \frac{1}{3}$$

을 구할 수 있다. 그러니까 해는

$$D_5(\xi) = \left(1 + \frac{\xi^2}{3}\right)^{-\frac{1}{2}} \qquad\qquad (3\text{-}2\text{-}45)$$

이 된다. 이 경우는 $\xi_1 \to \infty$라는 것을 알 수 있다. D_5의 그래프는 다음과 같다.

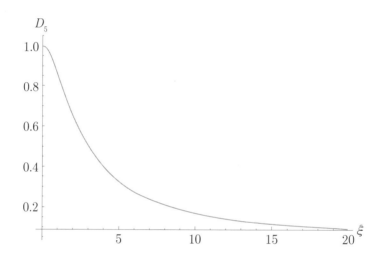

에딩턴의 별 모형_복사 압력을 추가하다

정교수   레인-엠덴 방정식을 써서 제대로 된 별의 모형을 만든 사람은 영국의 물리학자 에딩턴이야. 그의 연구를 알아보세.

에딩턴(Sir Arthur Stanley Eddington, 1882~1944)

에딩턴은 1882년 영국 켄들에서 태어났다. 1884년 영국을 휩쓴 장
티푸스로 아버지가 사망해, 그의 어머니는 적은 수입으로 에딩턴과
여동생을 양육해야 했다.

어릴 때부터 수학을 잘했던 에딩턴은 1898년에 맨체스터에 있는
오언스 칼리지(나중에 맨체스터 대학이 됨)에 입학해 물리학을 공부
했다. 1902년 그는 물리학과를 수석으로 졸업했다.

이후 케임브리지 대학 트리니티 칼리지의 대학원 과정에 진학해
1905년에 석사 학위를 받고, 캐번디시 연구소에서 열전자 방출에 대
해 연구했다.

1906년 1월 에딩턴은 왕립 그리니치 천문대의 왕립 천문학자 수석
조수가 되었다. 1914년에는 천문대 책임자로 임명되었다.

에딩턴은 천문 관측 외에도 이론에 관심이 많았는데, 특히 별의 내
부에 대한 이론을 만드는 일에 주의를 기울였다. 그는 1916년에 세
페이드 변광성을 연구하다가 레인-엠덴 방정식에 흥미를 보였다. 레

세상에서 가장 쉬운 과학 수업 별의 물리학

인-엠덴 방정식은 별 내부의 기체 압력과 중력이 평형을 이룬다는 모형이다. 에딩턴은 레인-엠덴의 모형에서 별 내부의 복사 압력을 추가해야 한다고 생각했다.

물리군 복사 압력이 뭐예요?

정교수 질량을 가진 기체 분자에 의한 압력을 기체 압력이라 하고, 질량이 0인 빛에 의한 압력을 복사 압력이라고 하네. 빛은 파동으로 해석하면 전자기파이고 입자로 해석하면 광자일세.

전자기파가 압력을 만드는 것은 1862년 전자기 방정식을 만든 맥스웰에 의해 알려졌다. 복사 압력의 첫 실험적인 관측은 1900년 레베데프(Pyotr Nikolaevich Lebedev, 1866~1912)와 1901년 니컬스(Ernest Fox Nichols, 1869~1924), 헐(Gordon Ferrie Hull, 1870~1956)에 의해 이루어졌다.

먼저 기체 압력을 생각해 보자. 에딩턴은 별 속의 기체를 이상기체로 가정했다. 기체의 입자 수를 N이라고 하면 이상기체 방정식으로부터 기체 압력 p_g는

$$p_g V = N k_B T \qquad\qquad (3\text{-}3\text{-}1)$$

를 만족한다. 여기서 V는 별의 부피이고 k_B는 볼츠만 상수이다. 입자 수 밀도 $n = \dfrac{N}{V}$ 을 도입하면

$$p_g = nk_BT \qquad\qquad (3\text{-}3\text{-}2)$$

가 된다.

별 속에는 서로 다른 질량을 갖는 기체들이 섞여 있다. 이때 별의 질량 M을 입자 수 N으로 나눈 것을 기체 입자의 평균 질량이라 하고 $\overline{m}$ 으로 쓴다.

$$\overline{m} = \frac{M}{N} \qquad\qquad (3\text{-}3\text{-}3)$$

별의 밀도 ρ는 질량을 부피로 나눈 값으로

$$\rho = \frac{M}{V} \qquad\qquad (3\text{-}3\text{-}4)$$

이 된다. 식 (3-3-3)과 (3-3-4)로부터

$$\rho = n\overline{m} \qquad\qquad (3\text{-}3\text{-}5)$$

을 얻을 수 있다. 이제 평균 분자량 μ를 다음과 같이 정의하자.

$$\mu = \frac{\overline{m}}{m_p}$$

여기서 m_p는 수소 원자의 질량[8]으로

8] 수소 원자는 수소의 원자핵(양성자 하나로 구성)과 전자 한 개로 이루어져 있으나 전자의 질량은 양성자의 질량에 비해 너무 작으므로 수소 원자의 질량은 양성자의 질량과 거의 같다.

$$m_p = 1.673532499 \times 10^{-27}(\mathrm{kg})$$

이다. 따라서 기체 압력은

$$p_g = \frac{\rho k_B T}{\mu m_p} \tag{3-3-6}$$

가 된다.

물리군   복사 압력은 어떻게 구하나요?

정교수   별 속의 광자들이 만드는 압력이 복사 압력이야.

물리군   질량이 없는 광자가 어떻게 압력을 만들죠?

정교수   뉴턴 역학으로는 설명이 안 되지. 하지만 아인슈타인의 특수
상대성이론을 생각하면 가능해.

진동수가 ν인 광자 하나는 에너지

$$\epsilon = h\nu$$

를 가진다. 이때 h는 플랑크 상수이다. 미국의 물리학자 콤프턴은 특
수상대성이론을 이용해 광자가 가지는 운동량이

$$\frac{\epsilon}{c} = \frac{h\nu}{c}$$

가 된다는 것을 알아냈다.

이제 다음 그림을 보자. 별을 반지름이 R인 공 모양으로 생각하고,

별 속의 광자가 구면의 아주 작은 넓이 요소 ΔA에 부딪쳐 압력을 만드는 과정을 상상할 것이다. 여기서 ΔA가 있는 곳의 편각은 θ이고 방위각은 ϕ이다.

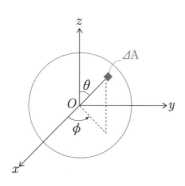

별 속의 광자들은 이리저리 움직이면서 구면 벽과 충돌한다. 이로 인해 복사 압력이 생긴다. ΔA가 너무 작기 때문에[9] 이 구면의 거의 xy 평면과 나란한(z축과 수직인) 면으로 볼 수 있다. 이때 z축 방향으로의 아주 작은 길이 요소 Δz를 생각하고, 다음과 같이 광자가 ΔA에 충돌하여 반사되는 경우를 가정하자.

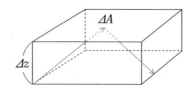

---

9) 거의 0에 가까운 넓이라고 생각하자.

세상에서 가장 쉬운 과학 수업 별의 물리학

그림에서 부피를 V라고 하면

$$V = \Delta A \Delta z$$

가 된다. 이 그림을 옆에서 보면 다음과 같다.

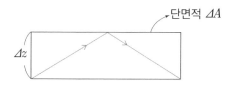

ΔA가 xy평면과 나란하다고 했으므로 운동량의 수직 성분(z성분) 만이 압력을 준다. 즉,

$$(\text{압력}) = \frac{(\text{힘의 } z\text{성분})}{(\text{넓이})} = \frac{(\text{운동량의 } z\text{성분의 변화량})}{(\text{시간}) \times (\text{넓이})}$$

이다.

ΔA에 입사하는 운동량의 z성분은

$$\frac{\epsilon}{c} \cos\theta$$

이고, 완전탄성충돌을 가정하면 ΔA에서 반사되는 운동량의 z성분은

$$-\frac{\epsilon}{c} \cos\theta$$

가 된다. 그러므로 운동량의 z성분의 변화량 크기는

$$2\frac{\epsilon}{c}\cos\theta$$

이다. 광자가 아랫면에서 윗면으로 움직여 반사된 후, 다시 아랫면으로 돌아오면서 윗면에 압력을 작용하는 경우를 생각하자. 이때 광자가 움직인 거리는

$$2\frac{\Delta z}{\cos\theta}$$

이므로 걸린 시간을 Δt라고 하면

$$2\frac{\Delta z}{\cos\theta} = c\,\Delta t$$

가 된다.

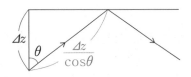

만일 부피가 V인 곳에서 면 ΔA에 충돌하는 광자의 수를 N'이라고 하면, 편각은 θ이고 방위각은 ϕ인 곳에 있는 ΔA에 작용하는 압력은

세상에서 가장 쉬운 과학 수업 별의 물리학

$$(\text{압력}) = \frac{N' \times 2\dfrac{\epsilon}{c}\cos\theta}{\dfrac{2\varDelta z}{c\cos\theta} \times A}$$

$$= \frac{N'\epsilon}{V}\cos^2\theta$$

이다. $\dfrac{N'\epsilon}{V}$ 은 부피 V 속의 광자의 총에너지를 부피로 나눈 값이므로 광자의 복사 에너지 밀도 u 이다. 즉,

$$(\text{압력}) = u\cos^2\theta$$

가 된다.

우리가 구한 압력은 구면의 한 점에 있는 아주 작은 넓이에 대한 복사 압력을 계산한 것이다. 구면 전체의 복사 압력을 구하려면 구면에 있는 모든 점에 대한 $u\cos^2\theta$의 평균을 계산해야 한다. 다시 말해 복사 압력을 p_r 라고 하면

$$p_r = u<\cos^2\theta>$$

이다. 여기서

$$<\cos^2\theta> = \frac{\displaystyle\iint \cos^2\theta dA}{\displaystyle\iint dA}$$

이고,

$$dA = (Rd\theta) \times (R\sin\theta d\phi) = R^2\sin\theta d\theta d\phi$$

이다.

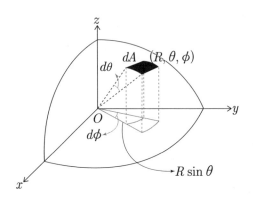

$$\iint \cos^2\theta dA = R^2 \int_0^\pi d\theta \cos^2\theta \sin\theta \int_0^{2\pi} d\phi$$

$$= \frac{1}{3} \times 4\pi R^2$$

이고,

$$\iint dA = 4\pi R^2$$

이므로

$$<\cos^2\theta> = \frac{1}{3}$$

이다. 따라서 복사 압력은 다음과 같다.

$$p_r = \frac{1}{3}u \qquad\qquad (3\text{-}3\text{-}7)$$

1879년에 오스트리아의 슈테판은 광자의 복사 에너지 밀도가 온도에만 의존하고, 온도의 네제곱에 비례한다는 것을 실험을 통해 알아냈다.

$$u = aT^4 \qquad\qquad (3\text{-}3\text{-}8)$$

즉, 온도가 T인 별의 복사 압력은 다음과 같다.

$$p_r = \frac{1}{3}aT^4 \qquad\qquad (3\text{-}3\text{-}9)$$

물리군 기체 압력과 복사 압력의 합이 중력에 대항하여 별의 모양을 유지하는 압력인가요?

정교수 맞아. 별의 압력을 p라고 하면

$$p = p_g + p_r \qquad\qquad (3\text{-}3\text{-}10)$$

라네. 만일 기체 압력이

$$p_g = \beta p \qquad\qquad (3\text{-}3\text{-}11)$$

이면

$$p_r = (1 - \beta)p \qquad\qquad (3\text{-}3\text{-}12)$$

일세. 식 (3-3-6)과 (3-3-11)을 이용하면

$$\frac{1}{3} a \left(\frac{\beta p \mu m_p}{\rho k_B} \right)^4 = (1 - \beta) p \qquad (3\text{-}3\text{-}13)$$

가 되네. 이 식으로부터

$$p = K \rho^{\frac{4}{3}} \qquad (3\text{-}3\text{-}14)$$

이고

$$K = \left[\frac{3(1 - \beta)}{a} \right]^{\frac{1}{3}} \left(\frac{k_B}{\beta \mu m_p} \right)^{\frac{4}{3}}$$

이 되지. 그러므로 복사 에너지를 고려하는 경우는 $n = 3$인 레인-엠덴 방정식과 관계된 것을 알 수 있다네.

네 번째 만남
·
별의 탄생과 진화

하버드 대학의 여성 천문학자들과 별의 분광형_별 연구의 역사

정교수　별의 탄생과 진화를 알기 위해서는 일단 별에 대한 연구의 역사를 살펴봐야 하네. 별이 수소와 헬륨으로 이루어진 것을 처음으로 밝혀낸 사람은 프라운호퍼(Joseph von Fraunhofer, 1787~1826)일세.

　1814년 프라운호퍼는 자신이 만든 분광기를 이용해 태양에서 오는 빛의 스펙트럼을 조사했다. 그 과정에서 스펙트럼에 검은 선이 존재하는 것을 알아냈다. 이는 태양을 이루는 원자들이 특정한 파장의 빛을 흡수하기 때문에 지구에 그 빛이 오지 않아 생기는 것이다. 이로써 태양이 어떤 원자들로 이루어져 있는지를 알게 되었다.

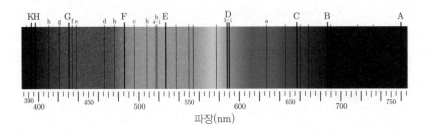

　태양의 표면은 대부분 수소(질량의 약 74%, 부피의 92%)와 헬륨(질량의 약 24~25%, 부피의 7%), 그 밖에 철을 비롯한 니켈, 산소, 규소, 황, 마그네슘, 탄소, 네온, 칼슘, 베릴륨, 크로뮴 등으로 구성되

　　　　　　　세상에서 가장 쉬운 과학 수업 별의 물리학

어 있다.

　헬륨(He)은 원자번호가 2로 수소 다음으로 가벼운 원소이며, 1868년 프랑스 천문학자 피에르 장센(Pierre Jules César Janssen)이 발견했다. 헬륨이라는 말은 그리스어로 태양을 뜻하는 헬리오스(Helios)에서 유래했다.

물리군　별은 온도가 제각기 다른가요?

정교수　물론이야. 미국의 여성 물리학자 캐넌이 별을 표면 온도에 따라 분류하는 일을 완성했지.

캐넌(Annie Jump Cannon, 1863~1941)

　캐넌은 1863년 미국 델라웨어주 도버에서 태어났다. 그의 아버지는 델라웨어주 조선업자이자 주 상원의원인 윌슨 캐넌이다. 캐넌은 어린 시절부터 어머니에게 별자리에 대해 배웠다. 두 사람은 다락방에서 보이는 별들을 식별하곤 했다.

도버에 있는 웨슬리 대학에서 수학을 공부한 캐넌은 1880년에 미국 최고의 여성 대학 중 하나인 매사추세츠의 웰즐리 대학에 들어가 물리학과 천문학을 배웠다. 1884년에 대학을 졸업한 그는 고향으로 돌아와 새로운 사진 기술을 개발했다.

1893년 캐넌은 성홍열에 걸려 청력을 상실하고 말았다.

1894년 어머니가 돌아가신 후 캐넌은 웰즐리 대학의 여교수인 화이팅에게 일자리를 부탁했다. 화이팅 교수는 그를 물리 강사로 채용했다. 캐넌은 강의 틈틈이 물리학 및 천문학 대학원 과정을 수강하면서 분광학을 공부했다.

화이팅(Sarah Frances Whiting, 1847~1927)

당시 하버드 대학은 여학생의 입학을 불허한 대신 여성들을 위해 래드클리프 대학을 세웠다. 캐넌은 이곳에 다니면서 천문학 연구를 지속했다. 비록 하버드 대학에 들어가지는 못했지만 래드클리프의

세상에서 가장 쉬운 과학 수업 별의 물리학

학생들도 하버드 대학 천문대(Harvard College Observatory)를 이용할 수 있었다. 1896년 하버드 대학 피커링 교수는 캐넌을 천문대 조수로 고용했다.

피커링(Edward Charles Pickering, 1846~1919)

하버드 대학 천문대

1896년에 캐넌은 피커링 교수의 별 분류 작업팀 중 여성들로만 구성되어 데이터를 정리하고 분석하는 역할을 하는 하버드 컴퓨터스(Harvard Computers)의 회원이 되었다.

하버드 컴퓨터스
(동그라미 안이 캐넌)

피커링 교수는 조수인 여성 물리학자 플레밍(Williamina Paton Stevens Fleming, 1857~1911)과 별들의 스펙트럼에서 수소에 해당하는 검은 선의 폭을 조사해 별을 분류하기 시작했다. 그들은 폭이 넓은 검은 선을 만드는 별을 'A형'이라고 부르는 식으로 별을 나누었다. 1901년 캐넌은 별을 표면 온도가 높은 것부터 차례로 O, B, A, F, G, K, M으로 분류했는데 이를 분광형이라 한다.

하버드 컴퓨터스의
작업 모습

세상에서 가장 쉬운 과학 수업 별의 물리학

이것을 손쉽게 외우는 방법은 다음과 같다.

Oh Be A Fine Girl Kiss Me

캐넌의 별 분류법은 하버드 대학에서 이루어졌으므로 하버드 분류법이라고 부른다. 캐넌은 일생 동안 395,000개의 별을 분류했다.

물리군   O, B, A, F, G, K, M은 각각 색깔이 다르겠군요.

정교수   물론이야. 표면 온도[10]에 따른 별의 색깔을 표로 정리해 볼까?

분광형	표면 온도	색깔
O	30,000도 이상	청색
B	10,000~30,000도	청백색
A	7,500~10,000도	백색
F	6,000~7,500도	황백색
G	5,200~6,000도	황색
K	3,700~5,200도	밝은 오렌지색
M	2,400~3,700도	주황색

물리군   태양은 어디에 속하나요?

정교수   태양은 표면 온도가 5,780도로 G형 별이야.

물리군   대표적인 O형 별은 뭐죠?

정교수   지구에서 제일 가까운 O형 별은 땅꾼자리 제타(Zeta

---

10) 좀 더 정확하게는 유효 온도이다. 유효 온도는 별에서 나오는 스펙트럼에서 세기가 가장 강한 빛의 파장에 대응하는 온도이다. 이것은 빈의 법칙으로부터 결정된다.

Ophiuchi)라네. 지구에서 366광
년 떨어져 있어. 표면 온도는 약
34,000도이고 질량은 태양의 약
20배, 지름은 약 8.5배 정도로 현
재 나이는 약 300만 년이지.

물리군　B형 별은 뭐가 있나요?

정교수　지구 근처의 대표적인 B
형 별로는 알골 A, 아케르나르,
아크룩스 등이 있다네.

땅꾼자리 제타

알골 A(출처: Dr Fabien Baron/Wikimedia
Commons)

아케르나르(출처: Pablo Carlos Budassi/
Wikimedia Commons)

아크룩스(출처: Alain r/Wikimedia Commons)

물리군　대표적인 A형 별은 뭔가요?

정교수　지구 근처의 A형 별로는 시리우스 A, 베가, 포말하우트, 알
타이르 등이 있지.

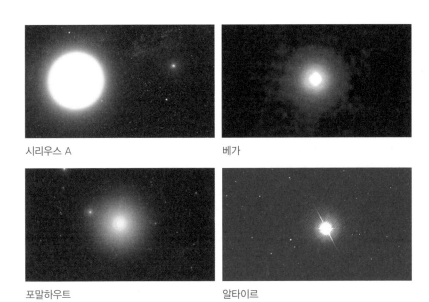

시리우스 A

베가

포말하우트

알타이르

물리군　F형 별에는 어떤 게 있나요?

정교수　폴라리스 Ab, B와 프로키온이 대표적인 F형 별이라네.

폴라리스 Ab, B

프로키온(왼쪽 위)

물리군   대표적인 G형 별은 뭔가요?

정교수   태양이 제일 대표적이고 그 외에 알파 켄타우리 A, 고래자리 타우 등이 있어.

알파 켄타우리 B(왼쪽)와
알파 켄타우리 A(오른쪽)
(출처: The plague/
Wikimedia Commons)

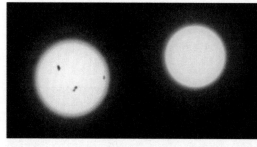

태양(왼쪽)과
고래자리 타우(오른쪽)
(출처: R.J. Hall/Wikimedia
Commons)

물리군   K형 별에는 어떤 게 있어요?

정교수   아르크투루스, 알데바란, 알파 켄타우리 B, 에리다누스자리 엡실론 등이 대표적인 K형 별이지.

세상에서 가장 쉬운 과학 수업 별의 물리학

아르크투루스

알데바란(출처: NASA, ESA 및 STScI)

에리다누스자리 엡실론(왼쪽)과
태양(오른쪽)
(출처: R.J. Hall/Wikimedia
Commons)

물리군　대표적인 M형 별에는 뭐가 있나요?

정교수　베텔게우스, 안타레스, 프록시마 켄타우리 등이 대표적인 M
형 별일세.

베텔게우스(출처: ESO/VLT)

안타레스(출처: ESO/K. Ohnaka)

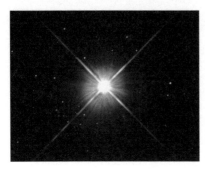

프록시마 켄타우리
(출처: ESA/Hubble)

H-R도의 발견_별의 밝기와 온도의 관계

정교수   별의 분광형이 알려진 후 덴마크의 헤르츠스프룽과 러셀이 별의 밝기와 온도의 관계를 연구했다네. 두 인물을 먼저 소개하기로 하지.

헤르츠스프룽(Ejnar Hertzsprung, 1873~1967)

세상에서 가장 쉬운 과학 수업 별의 물리학

헤르츠스프룽은 덴마크 프레데릭스베르에서 태어났다. 그는 코펜하겐 폴리테크닉 대학에서 화학공학을 공부했고 1898년에 졸업했다. 이후 러시아 상트페테르부르크에서 화학자로 2년을 보내고, 1901년 라이프치히 대학에서 1년 동안 광화학을 공부했다.

그의 아버지가 아마추어 천문학자였기 때문에 헤르츠스프룽은 어릴 때부터 천문학에 관심이 많았다. 그는 1902년에 프레데릭스베르에서 천문 관측을 시작했다. 그리고 몇 년 지나지 않아 유사한 분광형을 가진 별의 절대등급이 크게 다를 수 있음을 발견했다. 1909년에 그는 괴팅겐 천문대에서 직책을 맡았다.

러셀(Henry Norris Russell, 1877~1957)

러셀은 1877년 미국 뉴욕주 오이스터베이에서 목사의 아들로 태어났다. 그는 1895년 조지 스쿨을 졸업한 후 프린스턴 대학에서 천문학을 공부하여 1899년에 박사 학위를 받았다. 이후 1903년부터 1905년까지 케임브리지 천문대에서 연구 조교로 일했다. 1905년에는 다시 프린스턴 대학으로 돌아가 천문학 강사를 하다가 1908년에 교수가 되었다. 또한 1912년부터 1947년까지 프린스턴 대학 천문대(Princeton University Observatory)의 책임자를 지냈다.

물리군　두 사람이 공동 연구를 했나요?

정교수　그렇지는 않아. 독립적으로 같은 내용을 연구한 거지.

1911년 헤르츠스프룽은 별들의 절대등급과 온도 사이의 관계를 조사하기 시작했다. 이 연구는 1913년까지 계속됐다. 러셀 역시 독립적으로 같은 연구를 했고, 이를 1914년에 논문으로 발표했다. 두 사람의 연구 결과로부터 가로축을 별의 온도(또는 별의 분광형), 세로축을 별의 절대등급으로 택해 각각의 별을 배치한 표를 H-R도(H-R diagram)라고 부른다.

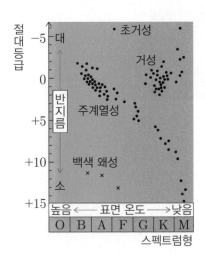

H-R도에서 세로축은 절대등급이고, 절대등급이 낮을수록 별은 밝아지므로 위로 올라갈수록 밝은 별을 나타낸다. 가로축은 별의 스펙트럼형(분광형)이므로 오른쪽으로 갈수록 표면 온도가 낮아진다.

헤르츠스프룽과 러셀은 관측된 별들의 분광형과 절대등급을 조사해 H-R도에 나타냈다. 관측된 별 중 80~90%가 왼쪽 위에서 오른쪽 아래로 내려가는 직선에 분포하는데, 이곳에 속하는 별들을 주계열성이라고 부른다. 또한 같은 분광형에서 주계열성보다 큰 별을 거성(giant star)[11], 그보다 훨씬 더 큰 별을 초거성이라고 한다. 주계열성보다 작은 별은 왜성(dwarf)[12]으로 부른다.

물리군   위로 올라갈수록 별이 커지는 이유는 뭐죠?

정교수   슈테판의 법칙 때문이야. 슈테판은 온도가 T이고 표면적이 A인 공 모양의 물체에서 나오는 복사 에너지는 표면적과 온도의 네제곱의 곱에 비례하는 것을 알아냈어. 별은 공 모양이므로 온도가 T이고 표면적이 A인 별의 복사 에너지도 슈테판의 법칙을 따르지. 그리고 복사 에너지가 클수록 별이 밝아지는데, 그 밝기를 나타내는 양을 별의 광도라고 한다네. 별의 광도를 I라고 하면

$$I \propto AT^4$$

이 돼. 광도가 크다는 것은 절대등급이 낮은 것을 뜻하지. 그러니까

---

11) 거인을 giant라고 한다.

12) 난쟁이별로도 부른다.

같은 온도에서 절대등급이 낮아지면(광도가 커지면) 별의 표면적이 커진다네. 같은 온도는 같은 분광형을 나타내니까, 같은 분광형일 때는 절대등급이 낮을수록 별이 커지는 거야.

물리군 그렇군요. 그럼 대표적인 거성은 뭔가요?

정교수 거성이라는 용어는 1905년 헤르츠스프룽이 처음 사용했어. 색깔에 따라 청색 거성, 적색 거성, 황색 거성 등으로 불리지. 대표적인 거성은 고래자리에 있는 미라(Mira)야. 미라는 적색 거성으로 지구에서 200~400광년 떨어져 있다네. 이 별은 1596년 8월 3일 독일 천문학자 다비트 파브리치우스(David Fabricius)가 발견했어.

미라

물리군 백색 왜성은 언제 발견되었나요?

정교수 최초로 발견된 백색 왜성은 40 에리다니(40 Eridani)라는 삼중성일세. 1783년 윌리엄 허셜이 발견했지.

물리군 삼중성이 뭐죠?

정교수 두 별이 가까이에 있는 경우를 이중성, 세 별이 가까이에 있는 경우를 삼중성으로 부른다네. 이때 가장 무거운 별을 주성, 다른 별을 동반성이라고 하지. 40 에리다니는 40 에리다니 A, 40 에리다니 B, 40 에리다니 C의 세 별로 이루어져 있어.

(40 에리다니 A의 질량) = (태양 질량)×0.78

(40 에리다니 B의 질량) = (태양 질량)×0.57

(40 에리다니 C의 질량) = (태양 질량)×0.20

물리군 40 에리다니 A가 주성이군요.

정교수 그렇지.

물리군 세 별 모두 백색 왜성인가요?

정교수 이 중 40 에리다니 B만이 백색 왜성일세.

백색 왜성 중에서 유명한 것은 시리우스 B(Sirius B)이다. 시리우스는 시리우스 A와 시리우스 B로 이루어진 이중성이다.

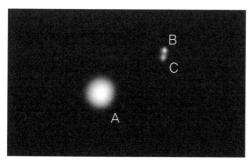

40 에리다니 A, B, C
(출처: Azhikerdude/Wikimedia Commons)

가운데 큰 원이 시리우스 A. 왼쪽 아래 작은 흰 점이 시리우스 B이다.

시리우스 A와 시리우스 B의 질량은 다음과 같다.

(시리우스 A의 질량) = (태양 질량)×2.063

(시리우스 B의 질량) = (태양 질량)×1.018

물리군   시리우스 A가 주성이군요.

정교수   맞아. 시리우스 A는 맨눈으로 볼 수 있는 별 중에서 가장 밝은 별이라네. 밝기는 태양의 25.4배 정도일세. 시리우스 A는 지구에서 약 8.59광년 떨어진 별로 태양계와 비교적 가까이에 있지. 시리우스 A는 주계열성이고 시리우스 B는 백색 왜성이야.

물리군   시리우스 B는 누가 발견했죠?

정교수   1844년 독일의 천문학자 베셀은 시리우스 A가 직선이 아니라 구불거리는 운동을 하는 것은 근처에 다른 별이 있어서 이 별의 인력의 영향을 받기 때문이라고 추론했어. 베셀이 생각한 별은 바로 시리우스 A의 동반성이었지.

1862년 미국의 클라크(Alvan Graham Clark, 1832~1897)는 시리우스 A 근처에서 시리우스 A의 만분의 1 정도 밝기인 어두운 별을 관측했다네. 이 별은 시리우스 B로 명명되었지. 시리우스 B의 반지름은 태양의 30분의 1, 질량은 태양의 0.96배, 밀도는 태양의 27000배 정도일세.

원시별 이론_별은 어떻게 탄생하는가

물리군 별은 어떻게 만들어지나요?

정교수 별이 만들어지는 곳은 성운이야. 성운(nebula)은 우주 공간에 분포한 성간물질이 좁은 지역에 밀집해 있는 것을 말한다네.

서기 150년경 프톨레마이오스는 저서 《알마게스트(Almagest)》에 큰곰자리와 사자자리 사이에서 별과 달리 흐릿하게 보이는 부분을 기록으로 남겼다. 아마도 이것이 처음으로 기록된 성운의 발견일 것이다.

1610년 페레스크(Nicolas-Claude Fabri de Peiresc, 1580~1637)가 망원경을 이용해 오리온성운을 최초로 발견했다.

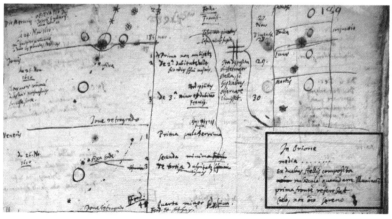

페레스크의 오리온성운 발견 노트

오리온성운

　1715년에 핼리(Edmond Halley, 1656~1742)는 6개의 성운 목록을 발표했다. 1746년에 슈조(Jean-Philippe de Cheseaux)는 20개의 성운 목록을, 1751년부터 1753년까지 라카유(Nicolas-Louis de Lacaille)는 42개의 성운 목록을, 1781년 샤를 메시에(Charles Messier)는 103개의 성운 목록을 만들었다.

　관측된 성운의 수는 허셜 남매(William Herschel과 그의 여동생 Caroline Herschel)의 노력으로 크게 증가했다. 1786년 그들은 1,000개의 새로운 성운과 성단[13] 목록을 출판했다.

　1864년 허긴스(Sir William Huggins, 1824~1910)는 70여 개의 성운에서 나오는 빛의 스펙트럼을 조사했다. 그리고 그중 약 1/3이 기체의 방출 스펙트럼(선스펙트럼으로 나타남)을 가지고 있음을 발견

---

13) 성운에서 같은 시기에 태어나 화학 조성과 나이가 거의 동일한 별들로 구성되어 있는 천체

했다. 나머지는 연속적인 스펙트럼을 보였으므로 성운 속의 별에서 나온 것으로 해석했다. 이를 통해 성운을 이루는 주요 물질은 수소나 헬륨과 같은 기체 상태라는 것이 알려졌다. 성운 자체는 스스로 빛을 내지 않지만 성운 속의 성간물질들이 별빛을 반사한 빛을 방출한다.

성운의 거의 대부분은 기체 상태인데 이것을 성간 기체라고 한다. 성운 속에도 아주 작은 고체 입자가 존재하는데 이를 우주먼지라고 부른다. 우주먼지는 1970년 미국의 전파망원경에 의해 처음 발견되었다. 우주먼지의 종류는 물, 철, 규소의 산화물 또는 메테인, 암모니아 같은 유기물질로 그 크기는 10만분의 1센티미터 정도이다. 이렇게 우주 공간에서 별과 별 사이에 존재하는 성간 기체와 우주먼지를 합쳐 성간물질이라고 부른다.

물리군 성간물질이 별을 만드나요?
정교수 맞아. 성간물질들이 만유인력으로 서로 끌어당기며 뭉쳐서 공 모양의 원시별을 만들지. 원시별은 수소 기체로 이루어진 별이야.

원시별(출처: ESO/ALMA)

물리군   성간물질이 얼마나 모여야 원시별이 되나요?

정교수   그 문제를 최초로 연구한 사람이 제임스 진스라네. 그의 연구를 자세히 살펴보세.

진스(Sir James Jeans, 1877~1946)

진스는 영국 랭커셔의 옴스커크에서 태어났다. 그의 아버지는 의회 특파원이자 작가였다. 진스는 케임브리지의 트리니티 칼리지에서 물리학을 공부했다. 그는 1901년 10월부터 트리니티 칼리지의 연구원으로 지내다 1904년에는 프린스턴 대학 응용수학 교수가 되었다. 그리고 1910년에 케임브리지 트리니티 칼리지로 돌아왔다.

레일리와 함께 흑체복사를 연구한 진스는 별의 진화를 포함하여 물리학의 많은 분야에 공헌했다. 그는 에딩턴과 더불어 영국의 우주론 연구 창시자이다.

1928년에 진스는 우주에서 물질이 지속적으로 생성된다는 가설을 바탕으로 정상 상태 우주론을 처음으로 주장했다. 1929년 은퇴한 후에는 일반 대중을 위해 《그들의 궤적에 빛나는 별들》(1931), 《우리 주변의 우주》(1929), 《공간과 시간을 통해》(1934), 《과학의 새로운 배경》(1933)과 같은 책을 집필해 과학의 대중화에 앞장섰다.

진스는 성운 속의 성간물질이 얼마나 뭉쳐야 원시별이 탄생하는가를 연구한 최초의 과학자이다. 그의 아이디어를 자세히 들여다보자.

성간물질이 공 모양을 이룬다고 가정하자. 성간물질은 기체 상태이므로 기체 분자의 운동 에너지를 가진다. 또한 성간물질은 질량을 가진 원자이므로 서로를 잡아당기는 중력이 작용한다. 진스는 비리얼 정리

$$2K + U = 0 \qquad\qquad (4\text{-}3\text{-}1)$$

을 떠올렸다. 여기서 K는 운동 에너지의 시간 평균, U는 퍼텐셜 에너지의 시간 평균을 의미한다.

식 (4-3-1)은 역학적 에너지 보존 법칙이 적용된 경우이며, 운동 에너지는 기체의 압력에 의한 에너지이다. 성간물질을 이상기체로 간주하면 운동 에너지는 온도 T일 때

$$K = \frac{3}{2}k_B NT \qquad\qquad (4\text{-}3\text{-}2)$$

로 주어진다. 이때 k_B는 볼츠만 상수이고 N은 성간물질의 입자 수이다.

한편 U는 중력에 의한 퍼텐셜 에너지이다. 그러므로 공 모양의 성간물질에 대한 평형 조건 (4-3-1)을 다시 쓰면

$$K = \frac{1}{2}|U|$$

(4-3-3)

이다. 만일 공 모양의 성간물질들이 뭉치면서 원시별을 이룬다면 중력에 의한 퍼텐셜 에너지가 기체의 압력에 의한 운동 에너지보다 커야 한다. 즉,

$$K < \frac{1}{2}|U|$$

(4-3-4)

를 만족해야 한다. 이것이 바로 진스가 찾은 원시별이 되기 위한 조건이다.

물리군 U는 어떻게 구하죠?

정교수 공 모양의 성간물질은 균일한 밀도를 갖는 질점으로 구성되었다고 볼 수 있어. 이때 각 질점 사이의 중력에 의한 퍼텐셜 에너지를 모두 더해야 공의 퍼텐셜 에너지가 구해지지.

원시별의 반지름을 R, 질량을 M이라고 하면 원시별의 밀도 ρ는

$$\rho = \frac{M}{\frac{4}{3}\pi R^3}$$

이 된다. 이제 다음 그림을 보자.

세상에서 가장 쉬운 과학 수업 별의 물리학

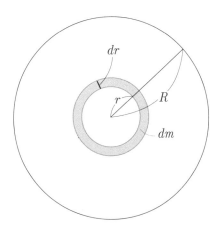

여기서 $r < R$이고 dm은 반지름이 r인 구를 에워싼 아주 얇은 구 껍질로 두께는 dr이다. 즉, dr는 거의 0에 가까울 정도로 매우 작은 값이다. 이때 구 껍질의 부피를 dV라고 하면

$$dV = \frac{4}{3}\pi (r+dr)^3 - \frac{4}{3}\pi r^3$$
$$= \frac{4}{3}\pi (r^3 + 3r^2dr + 3rdr^2 + dr^3) - \frac{4}{3}\pi r^3$$

이 된다. dr가 0에 가까울 정도로 너무 작아서 dr^2, dr^3의 값을 무시할 수 있으므로

$$dV = 4\pi r^2 dr$$

이다. 따라서 구 껍질의 질량 dm은

$$dm = \rho dV = 4\pi \rho r^2 dr$$

임을 알 수 있다.

구 껍질이 에워싼 부분의 질량은 반지름이 r인 구의 질량이다. 이 질량을 $M(r)$라고 할 때

$$M(r):M = \frac{4}{3}\pi r^3 : \frac{4}{3}\pi R^3$$

이고, 이를 정리하면 다음과 같다.

$$M(r) = \frac{r^3}{R^3}M$$

이때 구 껍질과 구 껍질이 에워싼 부분 사이의 중력에 의한 퍼텐셜 에너지를 dU라고 하면

$$dU = -G\frac{M(r)\,dm}{r} = -\frac{4\pi GM\rho}{R^3}r^4 dr$$

이다. 그러므로 반지름이 R, 질량이 M인 공 모양의 물체의 중력에 의한 퍼텐셜 에너지는

$$U = \int_{r=0}^{R} dU = -\int_{0}^{R}\frac{4\pi GM\rho}{R^3}r^4 dr = -\frac{3}{5}\frac{GM^2}{R}$$

이 된다. 즉, 진스 조건은

$$\frac{3}{5}\frac{GM^2}{R} > 2\times\frac{3}{2}Nk_BT$$

가 되므로 원시별의 질량 M은

$$M > \sqrt{\frac{5RNk_BT}{G}}$$

를 만족해야 한다. 이것이 바로 성간물질이 모여 원시별이 될 조건이며, 진스 조건이라고 부른다.

별이 빛나는 이유_별에서 일어나는 핵융합 반응

물리군   별은 어떻게 오랫동안 빛과 열을 낼 수 있죠?

정교수   그 문제를 처음 생각한 과학자는 영국의 에딩턴이라네.

   1920년경 에딩턴은 〈별의 내부 구성〉이라는 논문에서 별을 이루는 수소의 원자핵들이 고온에서 서로 달라붙는 과정이 별의 에너지를 만든다고 발표했다. 그는 높은 온도에서는 수소 원자핵 주위를 도는 전자가 큰 에너지를 얻어 원자핵으로부터 멀어지므로 수소는 전자를 잃어버린 양이온이 된다고 보았다. 양이온이 된 수소의 원자핵들이 높은 온도에서 달라붙으면서 헬륨의 원자핵을 만들어내는데, 이 과정에서 에너지가 발생해 별의 에너지가 된다는 것이 그의 생각이었다. 이렇게 높은 온도에서 원자핵들이 달라붙는 반응을 핵융합이라고 부른다.

1929년 앳킨슨(Robert Atkinson)과 하우테르만스(Fritz Houtermans)는 가벼운 원자핵들이 핵융합을 해 무거운 원소를 만들 때 많은 양의 에너지가 방출될 수 있다는 것을 확인했다.

물리군   에딩턴이 논문을 썼을 때는 중성자가 발견되기 전이네요.

정교수   맞아. 1932년 채드윅이 중성자를 발견한 후에 별 속에서 핵융합으로 원소들이 만들어지는 것을 알아낸 과학자는 독일의 한스 베테라네.

베테(Hans Albrecht Bethe, 1906~2005, 1967년 노벨 물리학상 수상, 사진 출처: LANL)

유대인인 베테는 1906년 스트라스부르[14]에서 태어났다. 그가 태어났을 때 아버지는 스트라스부르 대학의 학생이었고, 어머니는 스트라스부르 대학 교수의 딸이었다.

그의 아버지는 1912년 킬 대학의 생리학 연구소 교수 겸 소장직을

---

14) 당시에는 독일의 도시였지만 현재는 프랑스의 도시이다.

수락하여, 가족은 연구소 소장이 지내는 아파트로 이사했다. 베테는 7명의 다른 아이들과 함께 전문 교사로부터 사교육을 받았다. 그의 가족은 1915년에 아버지가 프랑크푸르트 대학의 새로운 생리학 연구소 소장으로 부임하면서 다시 이사했다.

12세의 베테와 그의 부모

베테는 독일 프랑크푸르트의 괴테 김나지움에 다녔다. 1924년 졸업 후에는 프랑크푸르트 대학에 진학했다. 그는 화학을 전공하기로 마음먹었지만 황산을 쏟아 실험복을 망가뜨릴 정도로 자신이 실험에 재주가 없는 것을 깨달았고, 이때부터 물리학에 흥미를 가지기 시작했다.

1926년 베테는 뮌헨 대학에 입학해 물리를 공부했다. 그리고 지도 교수인 조머펠트의 도움을 받아 결정의 전자 회절을 조사하는 내용으로 박사 학위를 받았다.

1930년 그는 영국 케임브리지 대학의 캐번디시 연구소에서 박사

후 과정을 밟기로 결정했고, 그곳에서 파울러(Ralph Fowler)의 지도를 받았다. 그 후 1931년에 로마에 있는 엔리코 페르미의 연구실로 가서 특정 1차원 양자 다체 모델의 고윳값과 고유벡터에 대한 정확한 솔루션을 찾는 베테 방법(Bethe ansatz)을 개발했다.

1932년 베테는 한스 가이거가 실험 물리학 교수로 있는 튀빙겐 대학의 조교수로 임명되었다. 하지만 나치 정부가 유대인 말살 정책을 시행하면서 대학에서 해고당했다.

그는 1933년 독일을 떠나 영국 맨체스터 대학에서 1년 동안 강사로 일했다. 당시 미국 코넬대 물리학과는 새로운 이론 물리학자를 물색 중이었고, 베테를 적합한 사람으로 여기게 되었다. 베테는 코넬대의 제안을 받아들여 1934년부터 코넬대 교수로 재직했다.

물리군　베테가 별의 탄생 이론을 만들었나요?
정교수　그렇다네. 별의 탄생 이론으로 노벨 물리학상을 받지.

이 책의 뒤에 첨부한 베테의 논문을 자세히 살펴보자.

원자핵을 줄여서 핵이라고도 하는데, 과학자들은 핵을 이루는 입자를 핵자(nucleon)라고 부른다. 각각의 원소는 서로 다른 개수의 양성자와 중성자를 갖는다. 양성자 수를 Z, 중성자 수를 N, 핵자수를 A로 나타내면

$$A = Z + N$$

이다. 전자의 질량은 양성자나 중성자에 비해 너무너무 작으므로 원자의 질량은 거의 원자핵의 질량과 같다. 원자의 질량을 원자량이라고 부르는데, 양성자와 중성자의 질량이 거의 비슷하므로 원자량은 양성자의 질량의 A배가 된다.

이제 별 속에서 일어나는 핵융합 과정을 설명하기 위해 몇 가지 입자의 기호를 정리하겠다.

p = H = 양성자(또는 수소의 원자핵)

n = 중성자

e = 전자

e^+ = 양전자

γ = 광자

ν = 뉴트리노

물리군　양전자는 뭐죠?

정교수　전자와 질량은 같은데 양의 전기를 띤 입자라네. 전자와 양전자의 전하량은 크기는 같고 부호는 반대일세. 1932년 미국의 앤더슨(Carl D. Anderson, 1936년 노벨 물리학상 수상)이 발견했어. 전자와 양전자가 만나면 빛으로 소멸한다는 것이 알려졌지. 이 과정은 다음과 같아.

$e + e^+ \rightarrow \gamma$

이것을 전자–양전자 소멸이라고 부르지.

물리군 뉴트리노는 뭔가요?

정교수 전자보다도 훨씬 가벼운(질량이 거의 0에 가까운) 입자야. 카원(Clyde Lorrain Cowan Jr, 1919~1974)과 라이너스(Frederick Reines, 1918~1998, 1995년 노벨 물리학상 수상)가 1956년에 발견했어. 뉴트리노는 전기를 띠지 않는 중성입자라네. 뉴트리노 발견으로 라이너스가 1995년에 노벨 물리학상을 받았지.

물리군 카원은 노벨상을 못 받았네요?

정교수 그해에 살아 있었다면 카원과 라이너스가 공동 수상했을 걸세. 하지만 노벨상은 죽은 사람에게는 수여하지 않는 원칙이 있거든.

다시 베테의 논문 속으로 들어가자. 우선 베타 붕괴 과정을 알아보자. 양성자가 중성자로 바뀌거나 중성자가 양성자로 바뀌는 과정을 베타 붕괴라고 하는데, 우리는 양성자가 중성자로 바뀌는 베타 붕괴 과정만 생각하기로 한다. 이 과정은 다음과 같은 반응식으로 나타낼 수 있다.

$$p \rightarrow n + e^{+} + \nu \qquad\qquad (4\text{-}4\text{-}1)$$

물리군 베타 붕괴 과정에서 양전자와 뉴트리노가 나오는군요.

정교수 맞아. 이제 수소 핵의 핵융합 과정을 자세히 설명해 볼게.

핵융합이란 두 개의 원자핵이 부딪혀 새로운 하나의 무거운 원자

핵으로 변환하는 반응이다. 이 과정에서 에너지가 발생한다. 별이 가진 에너지는 바로 핵융합에 의해 생긴 것이다. 이 에너지가 빛과 열을 낸다.

베테는 먼저 양성자-양성자 핵융합 반응을 세 종류로 설명했다.

제1종 양성자-양성자 핵융합 반응을 알아보자. 앞으로 우리는 수소의 원자핵을 H 또는 p로 쓸 것이다. 또한 어떤 원소 X의 핵자수가 N이면 이것을 X^N으로 쓴다.

두 개의 수소 핵이 만나면

$$H + H$$

이다. 높은 온도에서 양성자가 베타 붕괴 하므로

$$H + H \rightarrow p + n + e^+ + \nu \tag{4-4-2}$$

가 된다. 이때 $p + n$은 양성자 한 개와 중성자 한 개로 이루어진 원자핵을 의미한다. 이것을 중수소핵이라 부르고 D로 쓴다.[15] 식 (4-4-2)를 다시 쓰면

$$H + H \rightarrow D + e^+ + \nu \tag{4-4-3}$$

이다.[16] 수소 핵 두 개의 핵융합으로 만들어진 중수소핵과 수소 핵이

15] 베테는 논문에서 중수소핵을 D 또는 H²로 썼다.

16] 베테는 논문에서 뉴트리노를 고려하지 않았다.

다시 핵융합을 일으킨다. 그 과정은 다음과 같다.

$$H + D \rightarrow He^3 + \gamma \qquad (4\text{-}4\text{-}4)$$

여기서 He^3은 헬륨의 동위원소인 헬륨-3의 원자핵을 뜻하며, 양성자 두 개와 중성자 한 개로 이루어져 있다.

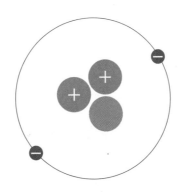

우리가 일상생활에서 보는 헬륨은 He^4로 그 원자핵은 양성자 두 개와 중성자 두 개로 이루어져 있다. 베테가 생각한 그다음 과정은 He^3 두 개의 핵융합이다. 이것을 식으로 나타내면 다음과 같다.

$$He^3 + He^3 \rightarrow He^4 + 2H \qquad (4\text{-}4\text{-}5)$$

세상에서 가장 쉬운 과학 수업 별의 물리학

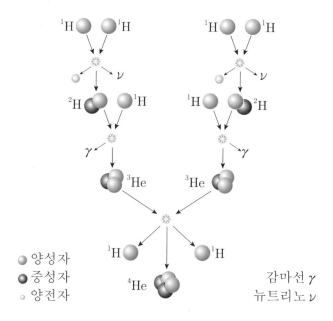

양성자
중성자
양전자

감마선 γ
뉴트리노 ν

물리군 제1종 양성자-양성자 핵융합 반응은 수소가 헬륨을 만드는 과정이군요.

정교수 그렇지. 이제 제2종 양성자-양성자 핵융합 반응을 설명할게. 제1종 양성자-양성자 핵융합 반응에서 만들어진 He^3과 He^4가 핵융합을 하면

$$He^3 + He^4 \rightarrow Be^7 \qquad (4\text{-}4\text{-}6)$$

이 되어, 원자번호 4인 베릴륨(Be)이 만들어지지. 그다음은 Be^7이 전자를 흡수해 리튬(Li)이 되는 과정이야.

$$Be^7 + e \rightarrow Li^7 + \nu + \gamma \qquad (4\text{-}4\text{-}7)$$

물리군 이 과정은 잘 이해가 안 돼요.

정교수 Be^7의 원자번호가 4이므로 양성자 수는 4가 돼. 그리고 핵자 수가 7이니까 중성자 수는 3이지. 이것을 다음과 같이 나타내겠네.

$$Be^7 = p + p + p + p + n + n + n$$

따라서

$$Be^7 + e = p + p + p + p + n + n + n + e$$

로 쓸 수 있어. 여기에서 베타 붕괴 과정의 식 (4-4-1)을 이용하면

$$p + e \rightarrow n + e^+ + \nu + e$$

이고, 전자-양전자 소멸에 따라

$$p + e \rightarrow n + \nu + \gamma$$

가 되지. 결국

$$Be^7 + e = p + p + p + n + n + n + n + \nu + \gamma$$

가 된다네. $p + p + p + n + n + n + n$은 양성자 수가 3이고 핵자수가 7인 원소의 원자핵이므로, 핵자수 7인 리튬 핵인 걸 알 수 있지.

$$p + p + p + n + n + n + n = Li^7$$

물리군 이제 이해가 되네요.

정교수   식 (4-4-7)의 다음 과정은

$$Li^7 + H \rightarrow 2He^4 \qquad\qquad (4\text{-}4\text{-}8)$$

일세.

물리군   그럼 제3종 양성자-양성자 핵융합 반응도 있나요?

정교수   베테가 생각한 그다음 핵융합 과정은 원자번호 5인 붕소(B)가 만들어지는 과정이야. 이것을 제3종 양성자-양성자 핵융합 반응이라고 부른다네. 식은 다음과 같아.

$$Be^7 + H \rightarrow B^8 + \gamma \qquad\qquad (4\text{-}4\text{-}9)$$

$$B^8 \rightarrow Be^8 + e^+ + \nu \qquad\qquad (4\text{-}4\text{-}10)$$

$$Be^8 \rightarrow 2He^4 \qquad\qquad (4\text{-}4\text{-}11)$$

베테는 이런 식으로 별 속에서 핵융합이 일어나면서 원소들이 만들어진다고 생각했지. 각각의 핵융합에서 에너지가 발생하는데 이것이 별의 에너지가 되는 거야.

물리군   별 속에서 모든 원소가 만들어지나요?

정교수   그렇지는 않아. 별 속에서 만들어지는 원소는 철까지라네.

물리군   왜 그런 거죠?

정교수   철 원자핵은 더 이상 핵융합 반응을 하지 않거든.

물리군   별이 갓 태어났을 때는 수소로만 이루어져 있다가 헬륨이 생기고 점점 무거운 원소들이 나타나는군요.

정교수   맞아. 그게 바로 별이 늙어가는 걸세.

별의 일생_별의 탄생부터 죽음까지

정교수 이제 별의 일생을 정리해 볼까?

 우주 공간에서 성간물질은 장소에 따라 차이가 난다. 성간물질이
희박한 곳이 있는가 하면 반대로 많이 모인 곳도 있다. 이들이 모여
우리 눈에 보이는 성운이 된다. 성운이란 성간물질로 이루어진 구름
이다. 영어로는 nebula라고 부른다.
 성운 속에서 성간물질은 만유인력으로 서로 끌어당긴다. 만유인력
은 거리의 제곱에 반비례하므로, 이들 사이의 거리가 가까워짐에 따
라 만유인력은 더 강해진다. 이렇게 강한 만유인력은 성간물질을 한
곳에 모이게 한다.

독수리 성운

용골자리 성운

세상에서 가장 쉬운 과학 수업 별의 물리학

성간물질이 어느 정도 모이면 서로의 만유인력에 의해 무리를 이루고, 모여든 성간물질이 점점 빨리 회전하며 중심 방향으로 중력에 의한 수축이 일어난다. 이때 중심부는 바깥쪽에 비해 훨씬 더 빨리 수축하며, 바깥쪽은 비교적 천천히 수축한다. 이 시기를 원시별이라고 하는데, 말하자면 갓 태어난 아기별이라고 볼 수 있다. 우주에는 아직도 원시별이 태어나고 있으며, 대표적으로 오리온성운에 원시별이 존재한다고 알려졌다.

오리온성운

모여든 성간물질이 회전하며 중심 쪽으로 수축하면 내부 온도는 올라간다. 이것이 바로 원시별이 뜨거운 이유이다.

태양의 원시별을 원시 태양이라고 부른다. 원시 태양은 지금의 태양에 비해 1000배나 밝고 크기도 100배 이상이었지만, 1000만 년 동안 수축하여 지금과 같은 안정된 모습을 유지하고 있다. 태양의 수명은 약 100억 년이고 현재 나이는 50억 년 정도이다.

이제 태양처럼 가벼운 별의 일생을 알아보자. 원시별로 태어나 핵융합 과정을 거쳐 중심부의 온도가 1000만 도가 되면, 중심부에서 수소가 헬륨으로 변하고 헬륨이 탄소로 바뀌는 핵융합 반응이 일어난다.

별의 중심부에서 수소가 사라지면 별의 중심부는 수축하면서 온도가 올라가기 시작한다. 그리고 바깥층에 남아 있는 수소들이 핵융합을 하면서 별이 붉게 부풀어 올라 적색 거성이 된다.

적색 거성은 표면 온도가 낮은 붉은 별임에도 불구하고 그 크기 때문에 밝게 보인다. 별은 질량과 관계없이 주계열성의 상태로 일생의 80~90%의 시간을 보낸다. 현재 태양은 주계열성이다.

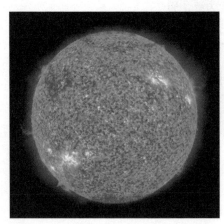

현재 주계열 상태인 태양

세상에서 가장 쉬운 과학 수업 별의 물리학

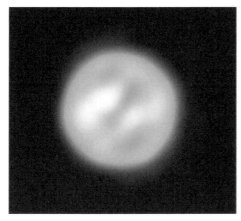

적색 초거성인 안타레스
(출처: ESO/K. Ohnaka)

가벼운 별들이 어떻게 죽음을 맞이하는지 살펴보자.

적색 거성의 중심부가 계속 수축하고 온도가 1억 도가 되면, 헬륨들이 핵융합을 해서 중심부에 탄소 핵이 만들어진다. 중심부의 헬륨이 모두 탄소로 바뀌면 가벼운 별은 더 이상 무거운 원소를 만들지 못한다. 그리고 바깥쪽에 있던 물질들이 외부로 방출되어 행성상 성운을 만들고, 중심부는 더욱 수축해서 백색 왜성이 된다.

백색 왜성에서는 더 이상 핵융합 반응이 일어나지 않는데, 이러한 별을 밀집성(compact star)이라고 부른다. 즉, 가벼운 별의 밀집성은 백색 왜성이다.

행성상 성운이라는 용어는 1780년대에 윌리엄 허셜이 고안했다. 망원경으로 들여다보았을 때 행성처럼 원반 모양의 상을 나타낸다고 하여 만들어진 말이다.

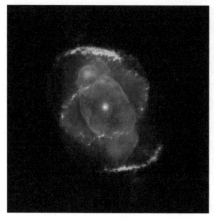

행성상 성운인 고양이 눈 성운

백색 왜성은 가벼운 별이 핵융합을 마치고 도달하는 천체를 말한다. 가벼운 별들은 중심부의 탄소가 핵융합을 일으키지 않으므로 탄소와 산소로 이루어진 핵만 남아 백색 왜성이 된다. 백색 왜성의 지름은 지구 크기와 비슷한 1만 킬로미터, 밀도는 1세제곱센티미터당 1톤, 표면 온도는 1만 도 정도이다.

이제 좀 더 무거운 별의 최후를 알아보자. 무거운 별에서는 탄소가 핵융합을 해 산소, 규소, 철 등이 만들어진다. 그러므로 가벼운 별에 비해 발생하는 에너지가 엄청 크다.

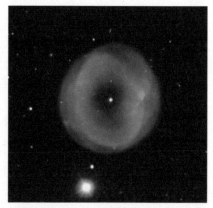

외부에 구름처럼 생긴 것이 행성상 성운, 가운데 파란 별이 백색 왜성이다.(출처: ESO)

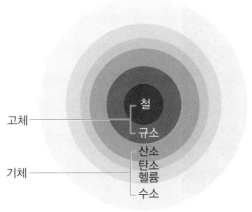

고체

기체

철
규소
산소
탄소
헬륨
수소

무거운 별의 내부 구조

철은 더 이상 핵융합을 하지 않는다. 따라서 공기가 빠진 풍선처럼 별 내부 압력이 약해져서 중력에 의해 중심 쪽으로 끌어당겨지는 중력수축이 시작된다. 이러한 중력수축은 빠른 속도로 진행되어 초신성 폭발이 일어난다. 남아 있는 별의 중심핵은 더욱 수축해 전자가 핵 속의 양성자와 만나 중성자가 되어 온통 중성자로만 구성된 별이 된다. 이것이 바로 중성자별이다. 즉, 무거운 별의 밀집성은 중성자별이다.

1987년 초신성 1987A(SN 1987A)가 대마젤란성운에 출현했다. 태양의 100배인 거성이 급격히 중력수축 하면서 발생한 이 폭발은 12등급의 어두운 별이 2개월 동안 2.9등성으로 밝아진 사실로부터 초신성 폭발로 입증되었다. 이 별은 15만 광년 떨어져 있으므로 이 초신성 폭발은 15만 년 전에 일어난 과거의 사건이다. 초신성 폭발은 1054년, 1572년, 1604년, 1987년에 관측되었을 만큼 희귀한 사건이고, 그 웅장한 광경 때문에 우주 쇼로 여겨진다.

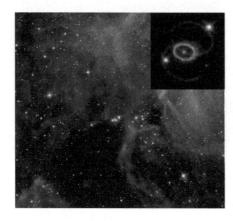

초신성 1987A(출처: ESA/Hubble)

초신성 폭발이 일어나면서 수많은 원시별이 탄생한다. 일부 과학
자들은 태양계가 초신성 폭발에 의해 만들어진 것으로 생각한다. 미
국 카네기 연구소의 앨런 보스(Alan Boss)와 샌드라 케이저(Sandra
Keiser)는 초신성 폭발의 충격파가 초기 태양 주위의 먼지 원반의 회
전을 일으켰다고 주장했다.

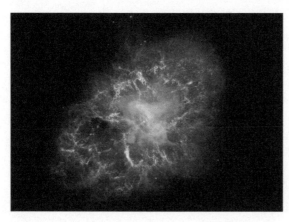

황소자리의 초신성 폭발로
만들어진 게성운

세상에서 가장 쉬운 과학 수업 별의 물리학

물리군 백색 왜성은 가벼운 별의 밀집성이고 중성자별은 무거운 별의 밀집성이라고 하셨잖아요? 구체적으로 질량이 어느 정도일 때 중성자별이 되나요?

정교수 별의 질량을 표시할 때는 태양 질량의 몇 배인가로 나타낸다네. 태양의 질량은 보통 $M_\odot$으로 나타내지. 밀집성의 질량이 $1.39M_\odot$보다 작으면 이 밀집성은 백색 왜성이고, $1.39M_\odot$에서 $3M_\odot$ 사이의 질량을 가진 밀집성은 중성자별이야. 이 기준을 처음 알아낸 사람은 인도 물리학자 찬드라세카르라네. 우리는 다음 장에서 찬드라세카르의 논문을 공부할 걸세.

물리군 관측된 가장 무거운 중성자별은 뭐죠?

정교수 J0740 + 6620이라는 별이야.

미국 웨스트버지니아의 전파망원경인 그린뱅크 망원경(GBT)으로 지구에서 약 4600광년 떨어진 이 중성자별을 발견했지.

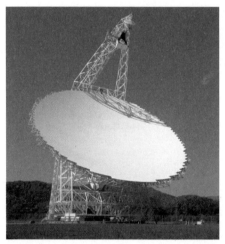

그린뱅크 망원경(출처: NRAO/AUI/NSF)

J0740＋6620은 자전 주기가 1000분의 1초로 엄청나게 빠르고 지름이 30킬로미터 정도라네. 질량은 $2.17M_\odot$으로 지구보다는 7만 배 이상 무겁지.

세상에서 가장 쉬운 과학 수업 별의 물리학

펄서_중성자별이 내는 규칙적인 신호

정교수 이번에는 펄서를 발견한 두 과학자를 알아볼까? 먼저 휴이시 일세.

휴이시(Antony Hewish, 1924~2021, 1974년 노벨 물리학상 수상, 사진 출처: Landesarchiv Baden-Württemberg, photo: Willy Pragher)

휴이시는 1924년 영국 콘월주 포위에서 은행가인 아버지의 세 아들 중 막내로 태어났다. 그는 대서양 연안의 뉴키에서 자라며 바다와 보트에 대한 사랑을 키웠다. 톤턴에 있는 킹스 칼리지에서 교육을 받은 그는 1942년에 케임브리지 대학에 진학했다.

대학에 다니면서 논문을 쓰고 있을 때 제2차 세계대전이 발발했다. 그는 영국 왕립 비행단(영국 공군)과 영국 원격통신 회사의 전쟁 지원을 위해 징집되었다. 그리고 영국 원격통신 회사에서 마틴 라일(Sir Martin Ryle, 1918~1984, 1974년 노벨 물리학상 수상)과 함께 공중 레이더 대응 측정 장치를 연구했다.

휴이시는 1946년에 케임브리지 대학으로 돌아와 1948년에 졸업하고, 즉시 캐번디시 연구소에 들어가 박사 학위를 취득했다. 그는 행성 간 섬광을 높은 시간 분해능으로 조사하는 데 사용할 대형 위상 배열 전파망원경의 건설을 제안했다. 그리고 1965년에 케임브리지 외곽의 MRAO(Mullard Radio Astronomy Observatory)에 자신이 설계한 행성 간 섬광 배열(Interplanetary Scintillation Array)을 건설하기 위한 자금을 확보했다. 이것은 1967년에 완성되었고 그의 박사 과정 학생 중 한 명인 조슬린 벨 버넬이 이 일을 도왔다.

MRAO
(출처: Rror/Wikimedia Commons)

세상에서 가장 쉬운 과학 수업 별의 물리학

벨 버넬(Dame Susan Jocelyn Bell Burnell, 1943~,
사진 출처: Roger W Haworth/Wikimedia Commons)

벨 버넬은 영국 북아일랜드 루건에서 태어났다. 그의 아버지는 아마(Armagh) 천문관 설계를 도운 건축가였다. 벨 버넬은 그곳을 종종 방문할 때마다 직원들로부터 천문학 분야에서 경력을 쌓도록 격려받았다. 그 또한 천문학에 관한 아버지의 책을 즐겨 읽었다.

1948년부터 1956년까지 벨 버넬은 루건 대학의 준비과에 다녔다. 당시 남학생들은 과학과 기술 과목을 배웠지만, 여학생들은 요리나 십자수 같은 과목을 공부해야 했다. 그는 부모님을 비롯한 여러 사람이 학교 정책에 이의를 제기한 후에야 과학을 공부할 수 있었다.

벨 버넬은 영국 요크에 있는 여학교인 마운트 스쿨에서 1961년에 중등 교육을 마쳤다. 다음으로 글래스고 대학에 입학하여 1965년에 물리학 학사 학위를 우등으로 따냈고, 그 후 케임브리지의 뉴홀에서 1969년에 박사 학위를 받았다.

최근 발견된 퀘이사를 연구하기 위해 벨 버넬은 휴이시 및 다른 사

람들과 협력하여 케임브리지 바로 외곽에 행성 간 섬광 배열을 건설
했다.

퀘이사

1967년 영국 케임브리지 대학의 천문학 박사 과정생이었던 벨 버
넬은 퀘이사를 연구하고자 전파망원경들을 설치했다. 조사 중 희한
하게 1.33초마다 규칙적인 신호가 오는 것을 발견했다. 그는 이 사
실을 지도 교수인 휴이시에게 알렸고, 두 사람은 이 신호가 외계인
이 보낸 것일지도 모른다고 생각했다. 그들은 이 수수께끼의 파동에
LGM-1(Little Green Men, 작고 푸른 사람)이라는 이름을 붙였다.
이는 외계 지성체를 희화하여 부르는 명칭이다.

하지만 한 달쯤 후 다른 곳에서 비슷한 천체를 또 하나 발견하면
서 이 신호가 외계인이 아닌 중성자별이 내는 것임을 깨달았다. 이후
LGM-1은 펄서(pulsar)라는 이름으로 불리게 되었다.

세상에서 가장 쉬운 과학 수업 별의 물리학

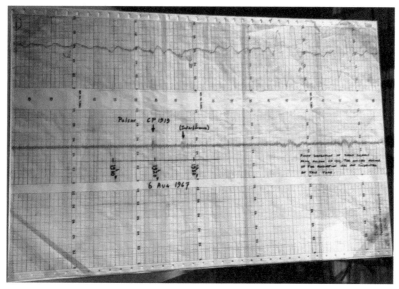

벨 버넬이 발견한 펄서에서 오는 전파 신호(출처: Billthom/Wikimedia Commons)

물리군   왜 이런 규칙적인 전파가 오는 거죠?

정교수   중성자별은 전파의 형태로 전자기파를 뿜으며 자전하는 별이야. 방출 빔이 지구를 향할 때만 펄서에서 오는 전파가 수신되지. 그런데 중성자별이 너무나 빠르게 자전하기 때문에 마치 회전하는 등대 불빛이 깜빡거리는 것처럼 전파가 왔다 안 왔다 하면서 규칙적인 신호를 보내는 거라네.

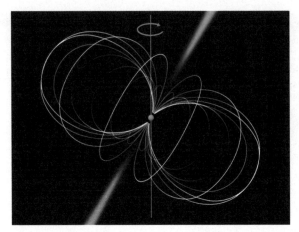

가운데는 중성자별, 주변의 곡선은 자기장선, 중성자별을 관통하는 푸른 광선은 방출 빔을 의미한다.(출처: Mysid, Jm smits/Wikimedia Commons)

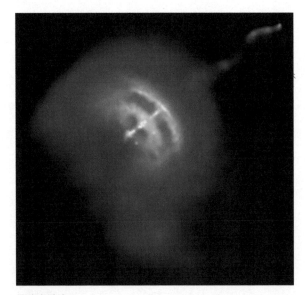

돛자리 펄서

세상에서 가장 쉬운 과학 수업 별의 물리학

블랙홀_빛조차 빠져나올 수 없는 천체

물리군　백색 왜성과 중성자별 외에 다른 밀집성이 있나요?

정교수　중성자별보다 더 무거워진 별은 다른 종류의 밀집성을 만들 수 있어. 밀집성의 질량이 $10M_\odot$보다 크면 중성자의 축퇴 압력이 밀집성을 지지하지 못해 중력붕괴로 블랙홀을 만들게 된다네.

물리군　왜 블랙홀이라고 부르죠?

정교수　블랙홀은 무거운 별의 종말로 엄청난 질량이 거의 한 점에 모여 있는 상황이야. 그러니까 중력이 무지무지 커지지. 블랙(검정)은 모든 빛을 흡수하는 성질을 가진다네. 엄청나게 큰 중력으로 인해 빛조차도 한번 들어가면 빠져나올 수 없는 천체이기 때문에 블랙홀이라고 부르는 걸세.

물리군　빛조차 탈출하지 못하는 블랙홀을 어떻게 찾아요?

정교수　우주의 별 중 절반 이상이 두 개의 별이 가까이 붙어 있는 연성이야. 두 별은 서로의 무게중심 주위를 회전하지. 이때 질량이 큰 별을 주성, 질량이 작은 별을 동반성이라고 부른다네.

　주성은 동반성보다 무겁기 때문에 더 밝게 빛나고 수소의 핵융합도 빨리 진행되어 별의 진화 속도가 더 빠르다. 태어난 지 약 1000만 년 정도 지나면 주성이 먼저 붉은 거성이 된다. 1200만 년 정도 지나면 주성은 마침내 초신성 폭발을 일으키고 별 바깥 부분의 가스는 초속 1만 킬로미터나 되는 속도로 우주 공간에 퍼져 나간다. 이때 주성의 밀집성이 무거우면 중심핵은 중력수축으로 블랙홀이 된다.

5000만 년 이상의 세월이 흐르면 동반성도 점점 커져 적색 거성이
된다. 그 팽창한 가스가 블랙홀이 되어 버린 주성의 강한 중력 때문에
블랙홀 안으로 빨려 들어간다. 이때 가스는 회전운동을 하므로 곧바
로 블랙홀로 빨려 들어가지는 않고 블랙홀 주위를 빠르게 회전하면
서 가스끼리 마찰을 되풀이한다. 마찰에 의해 가스는 뜨거워지고 온
도는 1000만 도에 도달해 강한 X선을 방출한다. 결국 블랙홀을 발견
하려면 X선을 방출하는 X선 별을 찾으면 된다.

현재 유럽의 엑소샛(EXOSAT)을 비롯한 3개의 X선 천문 위성이
X선 별을 관측하고 있다. 지금까지 관측된 X선 별은 천 개가 넘는다.

이들의 관측 결과에서 가장 유력한 블랙홀 후보로 등장한 것이 백조자리 X-1(Cygnus X-1)이다. 이 밖에도 대마젤란성운의 X-3도 블랙홀일 가능성이 매우 큰 것으로 알려져 있다.

2019년 4월에 공개된 실제 블랙홀의 모습(출처: ESO/EHT)

다섯 번째 만남

•

별의 죽음에 대한 연구

찬드라세카르의 생애_멈추지 않은 열정

정교수   이제 별의 죽음에 대한 연구로 노벨 물리학상을 받은 인도의 찬드라세카르를 소개하겠네.

찬드라세카르(Subrahmanyan Chandrasekhar, 1910~1995, 1983년 노벨 물리학상 수상)

찬드라세카르는 1910년, 영국령 인도 제국(현 파키스탄) 편자브주 라호르에서 태어났다. 아버지는 철도청 고급 관리였으며, 삼촌인 찬드라세카라 벵카타 라만은 라만 효과를 발견한 공로로 1930년에 노벨 물리학상을 수상한 물리학자였다.

찬드라세카르는 12살 때까지 집에서 개인 교습을 받았다. 아버지는 그에게 수학과 물리학을, 어머니는 타밀어를 가르쳤다. 12살에 첸나이로 이사한 그는 첸나이

라만(Chandrasekhara Venkata Raman, 1888~1970, 1930년 노벨 물리학상 수상)

힌두교 고등학교를 졸업하고, 첸나이 프레지던시 대학에서 학사, 석사 학위를 취득했다.

이후 영국의 인도 학생을 위한 장학금을 받아 1930년부터 케임브리지 대학 트리니티 칼리지에서 랠프 파울러의 지도 아래 수학하여, 1933년 박사 학위를 받았다.

찬드라세카르의 가장 대표적인 업적인 백색 왜성 연구는 대학원에 가기 위한 기나긴 배 여행 기간에 처음 시작하였다. 그는 태양보다 약 1.4배 이상 무거운 질량을 가진 별은 백색 왜성이 될 수 없음을 보였고, 이것을 바탕으로 여러 종류의 별의 마지막을 예측하였다.

그의 아이디어는 오랫동안 배척당했는데, 특히 천문학자 아서 에딩턴과의 논쟁은 유명하다. 당시 지도하는 입장에 있던 에딩턴은 찬드라세카르의 아이디어를 공개적으로 조롱하고, 그가 이론을 방어할 충분한 시간을 주지 않은 채 몰아붙였다고 한다. 찬드라세카르는 에딩턴과의 논쟁에 너무 기진맥진하는 바람에, 한때 다른 분야를 연구하려고 마음먹은 적도 있었다. 그러나 1930년대 후반이 지나면서 에딩턴을 제외한 대부분의 천체물리학자들은 그의 주장에 동조하기 시작한다.

박사 학위를 취득한 뒤, 찬드라세카르는 몇 년 동안 케임브리지 대학 트리니티 칼리지 및 하버드 대학에 머물렀다. 1953년에는 미국 시민권을 얻었다. 그리고 1995년 8월 1일 사망할 때까지 시카고 대학 여키스 천문대에서 교수직을 수행하며 교육 활동을 멈추지 않았다. 그는 생애 동안 50명이 넘는 대학원생들의 박사 연구를 지도하였다.

강의에 대한 정신이 투철했던 그는 240km 떨어진 시카고 대학까지 운전하여 다니며 강의를 했다. 이 강의를 수강한 학생 중 두 명(리정다오와 양전닝)은 노벨 물리학상을 받는다.

1999년에는 그의 이름을 딴 찬드라 X선 관측선이 우주로 쏘아 올려졌다.

찬드라 X선 관측선

찬드라세카르의 논문 속으로_백색 왜성의 평형 조건

정교수　그럼 찬드라세카르의 1931년 논문 속으로 들어가 보세. 내용에 조금 틀린 부분이 있어 바로잡으면서 설명하도록 하지. 논문 제목은 〈백색 왜성의 밀도〉라네.

찬드라세카르는 백색 왜성의 전체 압력 P를 복사 압력 p_r와 기체 압력 p_G의 합으로 나타냈다.

$$P = p_r + p_G \tag{5-2-1}$$

여기서

$$p_G = \beta P \tag{5-2-2}$$
$$p_r = (1 - \beta)P \tag{5-2-3}$$

로 두었다. 찬드라세카르는 백색 왜성이 죽은 별이므로 별의 복사 에너지가 너무 작아 무시할 수 있다고 보았다. 즉, 백색 왜성에서

$$\beta \approx 1$$

이라고 가정했다.

물리군   죽은 별이면 기체의 압력도 없는 거 아닌가요?

정교수   찬드라세카르는 전자의 축퇴 에너지가 압력을 만들어 백색 왜성을 공 모양으로 유지한다고 생각했어.

물리군   축퇴 에너지가 뭐죠?

정교수   백색 왜성 속의 전자들은 가능한 낮은 에너지 상태에 있으려는 성질이 있지.

예를 들어 별의 모양을 공 대신 한 변의 길이가 L인 상자로 가정해 보자. 이 속에 갇혀 있는 전자는 힘을 받지 않으므로 퍼텐셜 에너지가

0이다. 찬드라세카르는 상자 속의 전자를 양자역학으로 다루어야 한다고 보았다. 질량이 m인 전자의 슈뢰딩거 방정식은 다음과 같다.

$$-\frac{\hbar^2}{2m}\left(\frac{\partial^2}{\partial x^2}+\frac{\partial^2}{\partial y^2}+\frac{\partial^2}{\partial z^2}\right)\psi(x,y,z)=E\psi(x,y,z) \tag{5-2-4}$$

이 방정식을 풀면 전자의 파동함수는

$$\psi(x,y,z)=N\sin\left(\frac{n_x\pi}{L}x\right)\sin\left(\frac{n_y\pi}{L}y\right)\sin\left(\frac{n_z\pi}{L}z\right) \tag{5-2-5}$$

가 된다. 여기서

$$n_x,\,n_y,\,n_z=1,\,2,\,3,\,\cdots \tag{5-2-6}$$

이므로 전자의 에너지는 세 자연수 n_x, n_y, n_z에 의존한다. 이 에너지를 다음과 같이 쓰자.

$$E_{n_x,n_y,n_z}=\frac{\hbar^2\pi^2}{2mL^2}(n_x^2+n_y^2+n_z^2) \tag{5-2-7}$$

이때 $(n_x,\,n_y,\,n_z)$는 전자에 대한 양자 상태를 나타낸다.

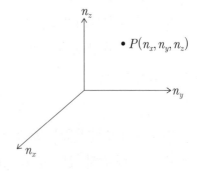

파울리의 배타원리에 의해 하나의 양자 상태에는 우회전 스핀 전자와 좌회전 스핀 전자가 가능하다. 따라서 하나의 양자 상태에 있을 수 있는 전자의 최대 수는 2개이다.

2차원의 경우 허용 가능한 상태는 다음과 같다.

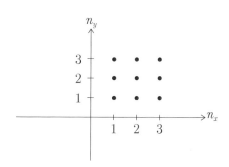

즉, 허용된 양자 상태는 제1사분면으로 국한된다.

이것을 3차원으로 확장하면 세 개의 축 n_x, n_y, n_z가 생긴다. 예를 들어 $\left(\frac{1}{2}, 1, 2\right)$는 허용 불가능한 점이고 $(1, 2, 4)$는 허용 가능한 점이다. 하나의 허용 가능 점에 전자 2개가 있을 수 있다. 그리고 n_x, n_y, n_z가 자연수이므로 허용된 양자 상태는 제1팔분공간으로 국한된다.

이제

$$n^2 = n_x^2 + n_y^2 + n_z^2 \tag{5-2-8}$$

으로 놓으면

$$n^2 = \frac{2mL^2}{\hbar^2 \pi^2} E(n) \qquad (5\text{-}2\text{-}9)$$

이 된다. 앞으로는 $E(n)$을 간단히 E라고 쓰자.

그러므로 주어진 에너지 E에 대해 허용 가능한 점은

$$n_x^2 + n_y^2 + n_z^2 \leq n^2$$

을 만족하는 모든 $(n_x,\, n_y,\, n_z)$이다.

별 속의 전자수는 너무너무 많으므로 반지름이 n인 구의 $\frac{1}{8}$ 쪽에 있는 $(n_x,\, n_y,\, n_z)$의 수는 구의 $\frac{1}{8}$ 쪽의 부피로 생각할 수 있다. 즉, 반지름이 n인 구의 $\frac{1}{8}$ 쪽에 있는 전자의 개수를 $N(n)$이라고 하면

$$N(n) = 2 \times \frac{1}{8} \times \frac{4}{3}\pi n^3 = \frac{1}{3}\pi n^3 \qquad (5\text{-}2\text{-}10)$$

이다. 여기서 2를 곱한 이유는 하나의 점에 두 개의 전자가 올 수 있기 때문이다.

식 (5-2-10)으로부터

$$\frac{dN}{dn} = \pi n^2 \qquad (5\text{-}2\text{-}11)$$

을 얻을 수 있다. 한편 식 (5-2-9)를 E로 미분하면

$$2n\,dn = \frac{2mL^2}{\hbar^2 \pi^2} dE \qquad (5\text{-}2\text{-}12)$$

이다. 따라서

$$dN = \frac{1}{2}\pi n \times 2ndn$$

$$= \frac{1}{2}\pi n \left(\frac{2mL^2}{\hbar^2\pi^2}\right)dE$$

$$= \frac{\sqrt{2m}\,mV}{\hbar^3\pi^2}\sqrt{E}\,dE \qquad\qquad (5\text{-}2\text{-}13)$$

가 된다. 여기서 $V = L^3$은 부피이다.

별 속에 N개의 전자가 있다고 하고 이때 허용되는 상태의 최대 에너지를 E_m으로 놓으면

$$N = \int_0^{E_m} dN = \frac{\pi}{3}\left(\frac{2mE_m}{\hbar^2\pi^2}\right)^{\frac{3}{2}}V \qquad\qquad (5\text{-}2\text{-}14)$$

가 되어, 전자의 수는 부피에 비례한다. 한편 E_m에 대응하는 n을 n_m이라고 하면

$$n_m^2 = \frac{2mL^2}{\hbar^2\pi^2}E_m \qquad\qquad (5\text{-}2\text{-}15)$$

이다. 이 식은

$$E_m = \frac{\pi^2\hbar^2}{2m}\left(\frac{3N}{\pi V}\right)^{\frac{2}{3}} \qquad\qquad (5\text{-}2\text{-}16)$$

으로 쓸 수 있다.

따라서 백색 왜성 속 전자들의 총에너지를 U라고 하면 다음과 같다.

$$U = \int_0^{n_m} E(n)\,dN$$

$$= \frac{\pi^2 \hbar^2}{2mL^2} \int_0^{n_m} n^2 \times \pi n^2\,dn$$

$$= \frac{\pi^3 \hbar^2}{2mL^2} \int_0^{n_m} n^4\,dn$$

$$= \frac{\pi^3 \hbar^2}{2mL^2} \frac{n_m^5}{5}$$

$$= \frac{\pi^3 \hbar^2}{10mL^2} \left(\frac{2mL^2}{\hbar^2 \pi^2} E_m \right)^{\frac{5}{2}}$$

$$= \frac{\pi^3 \hbar^2}{10mL^2} \left(\frac{3N}{\pi} \right)^{\frac{5}{3}}$$

$$= \frac{1}{10m} \hbar^2 \pi^{\frac{4}{3}} 3^{\frac{5}{3}} N^{\frac{5}{3}} V^{-\frac{2}{3}} \qquad (5\text{-}2\text{-}17)$$

이것을 축퇴 에너지라고 부르는데 바로 백색 왜성의 내부 에너지이다.

찬드라세카르는 백색 왜성에서 단열과정이 일어난다고 생각했다. 그러므로 식 (3-1-17)에 의해 축퇴 에너지에 의한 압력(축퇴 압력)은

$$p_G = -\frac{\partial U}{\partial V} = \frac{3}{5}\left(\frac{\pi^4}{3}\right)^{\frac{1}{3}}\frac{\hbar^2}{m}\left(\frac{N}{V}\right)^{\frac{5}{3}} \tag{5-2-18}$$

이 된다. 백색 왜성의 전자밀도(단위부피당 전자수)를

$$n = \frac{N}{V} \tag{5-2-19}$$

으로 정의하면

$$p_G = \frac{3}{5}\left(\frac{\pi^4}{3}\right)^{\frac{1}{3}}\frac{\hbar^2}{m}n^{\frac{5}{3}}$$

또는

$$p_G = \frac{\pi}{60}\frac{h^2}{m}\left(\frac{3n}{\pi}\right)^{\frac{5}{3}} \tag{5-2-20}$$

이다.

백색 왜성은 죽어가는 별이므로 복사 압력이 축퇴 압력보다 작아

$$P \approx p_G = \frac{\pi}{60}\frac{h^2}{m}\left(\frac{3n}{\pi}\right)^{\frac{5}{3}} \tag{5-2-21}$$

이 된다.

물리군 찬드라세카르의 논문에 나오는 식이 등장했네요. 논문에는

왜 이런 설명을 안 쓰는 거죠?

정교수 논문은 지식수준이 어느 정도 되는 사람들을 대상으로 집필하기 때문에 자세한 계산 과정은 적지 않아. 특히 이론 물리학 논문은 더 그렇지.

찬드라세카르는 전자수 밀도를 구하기 위해, 별 속에 들어 있는 수많은 종류의 원소를 고려했다. 별은 온도가 높기 때문에 전자들이 충분한 에너지를 얻어 핵의 영향에서 벗어날 수 있다. 따라서 남아 있는 원자핵은 양의 전기를 띤 이온이 된다.

즉, 별의 기체 압력은 이온에 의한 압력과 전자에 의한 압력의 합으로

$$p_G = p_{ion} + p_e \tag{5-2-22}$$

로 쓸 수 있다. 여기서 p_{ion}은 이온에 의한 압력, p_e는 전자에 의한 압력을 나타낸다.

핵 안의 이온의 종류는 수소, 헬륨 등으로 다양하다. 각각의 이온에 1, 2, 3, 4 등으로 이름을 붙이자. 이온 i에 의한 압력을 p_i라고 하면

$$p_{ion} = \sum_i p_i$$

이다. 별 속의 전자의 개수를 N_e, 이온 i의 개수를 N_i라고 할 때

세상에서 가장 쉬운 과학 수업 별의 물리학

$$p_i V = N_i k_B T$$

$$p_e V = N_e k_B T$$

로 쓸 수 있다.

이제 다음과 같이 놓자.

$$n_i = (이온 i의 개수 밀도) = \frac{N_i}{V}$$

$$n = (전자수 밀도) = \frac{N_e}{V}$$

따라서 다음과 같이 나타낼 수 있다.

$$P = \left(\sum_i n_i + n\right) k_B T \qquad (5-2-23)$$

별의 총질량을 M이라고 하면 별의 밀도는

$$\rho = \frac{M}{V}$$

이다. 이온 i의 핵자수를 A_i라고 할 때, 이온 i 한 개의 질량은 핵자수와 양성자의 질량 m_p의 곱이므로

(이온 i 한 개의 질량) $= A_i m_p$

(이온 i의 전체 질량) $= N_i A_i m_p$

이다. 전자의 질량은 이온의 질량에 비해 무시할 수 있으므로

$$M = \sum_i N_i A_i m_p$$

가 된다. 이때

$$X_i = \frac{N_i A_i m_p}{M} = \frac{N_i A_i m_p}{\rho V} = \frac{n_i A_i m_p}{\rho} \qquad (5\text{-}2\text{-}24)$$

로 놓으면

$$\sum_i X_i = 1$$

이다. X_i는 별의 전체 질량 중 이온 i들의 질량이 차지하는 비율을 뜻하며

$$n_{ion} = \sum_i \frac{X_i \rho}{A_i m_p}$$

가 된다. 여기서

$$\frac{1}{\mu_{ion}} = \sum_i \frac{X_i}{A_i}$$

로 놓으면

$$n_{ion} = \frac{1}{\mu_{ion}} \frac{\rho}{m_p}$$

또는

세상에서 가장 쉬운 과학 수업 별의 물리학

$$P_{ion} = n_{ion}kT = \frac{1}{\mu_{ion}}\frac{\rho}{m_p}kT \tag{5-2-25}$$

가 된다.

이제 이온 i의 원자 번호를 Z_i라고 하면, 이온 i에 대응하는 전자수 밀도는

$$Z_i n_i$$

이므로 전자수 밀도 n은

$$n = \sum_i Z_i n_i$$

$$= \sum_i Z_i \frac{X_i \rho}{A_i m_p}$$

임을 알 수 있다. 이것을

$$n = \frac{1}{\mu_e}\frac{\rho}{m_p}$$

로 정의하면

$$\mu_e = \left(\sum_i \frac{Z_i X_i}{A_i}\right)^{-1}$$

이다. 그러므로

$$P \approx p_G = \frac{(3\pi^2)^{\frac{2}{3}} \hbar^2}{5m} \left(\frac{\rho}{\mu_e m_p} \right)^{\frac{5}{3}}$$

이다. 즉,

$$P = K\rho^{\frac{5}{3}} \qquad\qquad (5\text{-}2\text{-}26)$$

이 된다. 여기서

$$K = \frac{\pi h^2}{60m} \left(\frac{3}{\pi m_p} \right)^{\frac{5}{3}} \frac{1}{\mu_e^{\frac{5}{3}}} \qquad\qquad (5\text{-}2\text{-}27)$$

이다.

찬드라세카르는 백색 왜성의 밀도가 ρ로 일정한 경우를 가정하고 중력에 의한 압력 p를 구했다. 세 번째 만남에서 다룬 공식

$$\frac{dp}{dr} = -\frac{GM(r)\rho(r)}{r^2}$$

에서

$$M(r) = \rho \times \frac{4}{3}\pi r^3$$

임을 알 수 있다. 따라서

$$\frac{dp}{dr} = -\frac{4\pi}{3}G\rho^2 r$$

세상에서 가장 쉬운 과학 수업 별의 물리학

이고, 이것을 적분하면 중력에 의한 압력은

$$p(r) = -\frac{2\pi}{3}G\rho^2 r^2 + C$$

가 된다. 백색 왜성의 표면에서 중력에 의한 압력이 0이라고 하면

$$p(R) = 0$$

이다. 그러므로

$$C = \frac{2\pi}{3}G\rho^2 R^2$$

이고, 중력에 의한 압력은

$$p(r) = \frac{2\pi}{3}G\rho^2(R^2 - r^2) \qquad\qquad (5\text{-}2\text{-}28)$$

이 된다.

찬드라세카르는 백색 왜성의 평형 조건은 별의 중심부에서 중력에 의한 압력과 전자의 축퇴 압력이 같은 것이라고 생각했다. 별의 중심에서 중력에 의한 압력은

$$p(0) = \frac{2\pi}{3}G\rho^2 R^2$$

이므로

$$\frac{2\pi}{3}G\rho^2 R^2 = \frac{(3\pi^2)^{\frac{2}{3}}\hbar^2}{5m}\left(\frac{\rho}{\mu_e m_p}\right)^{\frac{5}{3}}$$

이다. 따라서

$$\rho R^6 = (일정)$$

하다. 그리고

$$\rho \times \frac{4}{3}\pi R^3 = (백색 왜성의 질량) = M$$

으로부터

$$MR^3 = (일정) \tag{5-2-29}$$

함을 알 수 있다. 이것이 바로 찬드라세카르가 발견한 백색 왜성의 평형 조건이다. 즉, 백색 왜성이 평형 상태에 있으려면 질량이 반지름의 세제곱에 반비례해야 한다.

물리군 찬드라세카르는 백색 왜성이 평형을 유지하기 위한 질량과 반지름의 관계를 발견한 거군요.

정교수 맞아. 이 업적으로 찬드라세카르는 노벨 물리학상을 받았지.

만남에 덧붙여

on the Theoretical Temperature of the Sun.

J. H. Lane

ART. IX. — *On the Theoretical Temperature of the Sun; under the Hypothesis of a Gaseous Mass maintaining its Volume by its Internal Heat, and depending on the Laws of Gases as known to Terrestrial Experiment;* by J. HOMER LANE, Washington, D. C.

[Read before the National Academy of Sciences at the session of April 13–16, 1869.]

MANY years have passed since the suggestion was thrown out by Helmholtz, and afterwards by others, that the present volume of the sun is maintained by his internal heat, and may become less in time. Upon this hypothesis it was proposed to account for the renewal of the heat radiated from the sun, by means of the mechanical power of the sun's mass descending toward his center. Calculations made by Prof. Pierce, and I believe by others, have shown that this provides a supply of heat far greater than it is possible to attribute to the meteoric theory of Prof. Wm. Thomson, which, I understand, has been abandoned by Prof. Thomson himself as not reconcilable with astronomical facts. Some years ago the question occurred to me in connection with this theory of Helmholtz whether the entire mass of the sun might not be a mixture of transparent gases, and whether Herschel's clouds might not arise from the precipitation of some of these gases, say carbon, near the surface, with their revaporization when fallen or carried into the hotter subjacent layers of atmosphere beneath; the circulation necessary for the play of this Espian theory being of course maintained by the constant disturbance of equilibrium due to the loss of

216 세상에서 가장 쉬운 과학 수업 별의 물리학

heat by radiation from the precipitated clouds. Prof. Espy's theory of storms I first became acquainted with more than twenty years ago from lectures delivered by himself, and, original as I suppose it to be, and well supported as it is in the phenomena of terrestrial meteorology, I have long thought that Prof. Espy's labors deserve a more general recognition than they have received abroad. It is not surprising, therefore, in a time when the constitution of the sun was exciting so much discussion, that the above suggestions should have occurred to myself before I became aware of the very similar, and in the main identical, views of Prof. Faye, put forth in the Comptes Rendus. I sought to determine how far such a supposed constitution of the sun could be made to connect with the laws of the gases as known to us in terrestrial experiments at common temperatures. Some calculations based upon conjectures of the highest temperature and least density thought supposable at the sun's photosphere led me to the conclusion that it was extremely difficult, if not impossible, to make out the connection in a credible manner. Nevertheless, I mentioned my ideas to Prof. Henry, Secretary of the Smithsonian Institution, when he immediately referred me to a number of the Comptes Rendus, then recently received, containing Faye's exposition of his theory. Of course nothing is further from my purpose than to make any kind of claim to any thing in that publication. After becoming acquainted with his labors I still regarded the theory as seriously lacking, in its physical or mechanical aspect, the direct support of confirmatory observations, and even as being subject to grave difficulty in that direction. In this attitude I allowed the subject to rest until my friend Dr. Craig, in charge of the Chemical Laboratory of the Surgeon General's office, without any knowledge of Faye's memoir, or of my own suggestions previously made to Prof. Henry and another scientific friend, fell upon the same ideas of the sun's constitution, availing himself, precisely as I had done, of Espy's theory of storms. Dr. Craig's ideas were communicated to a company of scientific gentlemen early last spring, and soon after, Prof. Newcomb, of the U. S. Naval Observatory, entered into a general survey of the nebular hypothesis. These communications of Dr. Craig and Prof. Newcomb led me to enter into a renewed examination of the mechanical embarrassment under which I had believed the theory to labor. Not any longer relying on my first rough estimate based on assumed high temperatures at the photosphere, the question was now inverted. Assuming the gaseous constitution, and assuming the laws expressed in Poisson's formulæ, known to govern the constitution of gases at common temperatures and densities, what shall we find to be the temperatures and densities corresponding to the observed volume of the sun supposing

it were composed of some known gas such as hydrogen, or supposing it to be composed of such a mixture of gases as would be
represented by common air. Pure hydrogen will, of course,
give us the lowest temperature of all known substances, under
the general hypothesis.

The question was resolved, and the results were communicated in graphical and numerical form in May or June last to two
or three scientific friends, but their publication has been delayed
by an unavoidable absence of several months from home.

Premising that the unit of density shall correspond to a unit
of mass in the cube of the unit of length, the unit of force to
the force of terrestrial gravity in the unit of mass, and the unit
of pressure or elasticity in the gas to the unit of force on a
surface equal to the square of the unit of length :

Let $r=$ the distance of an element of the sun's mass from the
sun's center,

$t=$ the temperature of the element,

$\sigma t=$ its atmospheric subtangent, referred to the force of
gravity at the earth's surface, or height of the column
of homogeneous gas, whose terrestrial gravitating force
would equal its elasticity,

$\varrho=$ its density, or mass of its unit volume,

$=$ force of terrestrial gravity in its unit volume,

$\varrho \sigma t=$ its elasticity, or elastic force per unit surface,

$R=$ the earth's radius,

$M=$ the earth's mass,

$m=$ the mass of the part of the sun's body contained in
the concentric sphere whose radius is r,

$\dfrac{M}{m}\dfrac{r^2}{R^2} \sigma t=$ the subtangent of the gas under its actual gravitating force in the sun.

The condition of equilibrium between the gravitating force
of a thin horizontal layer of gas whose thickness is dr, and the
difference of elastic force between its lower and upper surfaces,
is expressed by the equation,

$$d \cdot \varrho \sigma t = - \frac{m}{M} \frac{R^2}{r^2} \varrho \, d r.$$

Under the hypothesis that the law of Mariotte and the law
of Poisson prevail throughout the whole mass, and that this
mass is in convective equilibrium, we have

$$\sigma = \text{a constant}, \qquad\qquad (1)$$
$$t = t_1 \varrho^{k-1}, ^*$$

t_1 representing the value of t in the part of the mass where the
density is a unit.

The theoretical difficulties which, if the supply of solar heat

* k represents the ratio of the specific heat of a gas under constant pressure to
its specific heat under constant volume.

세상에서 가장 쉬운 과학 수업 별의 물리학

is to be kept up by the potential due to the mutual approach of the parts of the sun's mass consequent on the loss of heat by radiation, come in when we suppose a material departure from these laws of Mariotte and of Poisson at the extreme temperatures and pressures in the sun's body, or how far such difficulties intervene, will be considered further on.

By means of the constant value of σ, and the value of t given in (1), the above differential equation is transformed into

$$k \sigma t_1 \varrho^{k-2} d\varrho = -\frac{m}{M} \frac{R^2}{r^2} d r,$$

the integral of which gives

$$1 - \left(\frac{\varrho}{\varrho_0}\right)^{k-1} = \frac{k-1}{k} \frac{R^2}{\sigma M t_1 \varrho_0{}^{k-1}} \int_0^r \frac{m dr}{r_2}, \qquad (2)$$

in which ϱ_0 is the value of ϱ at the sun's center.

We have also

$$m = 4\pi \int_0^r \varrho r^2 dr = 4\pi \varrho_0 \int_0^r \frac{\varrho}{\varrho_0} r^2 dr. \qquad (3)$$

If now we put

$$r = \left(\frac{k \sigma M t_1}{4(k-1) R^2 \pi \varrho_0{}^{2-k}}\right)^{\frac{1}{2}} x, \qquad (4)$$

we shall have

$$m = 4\pi \varrho_0 \left(\frac{k \sigma M t_1}{4(k-1) R^2 \pi \varrho_0{}^{2-k}}\right)^{\frac{3}{2}} \mu, \qquad (5)$$

in which

$$\mu = \int_0^x \frac{\varrho}{\varrho_0} x^2 dx, \qquad (6)$$

and equation (2) becomes

$$1 - \left(\frac{\varrho}{\varrho_0}\right)^{k-1} = \int \frac{\mu dx}{x^2}. \qquad (7)$$

In equations (6) and (7) it is plain that upon the value of k alone depends: first the form of the curve that expresses the value of $\frac{\varrho}{\varrho_0}$ for each value of x; secondly, the value of the upper limit of x corresponding to $\frac{\varrho}{\varrho_0} = 0$; and thirdly, the corresponding value of μ. These limiting, or terminal, values of x and μ, cannot be found except by calculating the curve, for equations (6) and (7) seem incapable of being reduced to a complete general integral. But when these values have been found for any proposed value of k, they may be introduced once for all into equations (4) and (5), from which the values of ϱ_0, and of σt_1, are at once deduced.

I have made these calculations for two different assumed values of k, viz., $k = 1\cdot4$, which is near the experimental value it has in common air, and $k=1\frac{2}{3}$, which is the maximum possible value it can have in the light of Clausius' theory of the constitution of the gases. The calculation of the curve of $\frac{\varrho}{\varrho_0}$, or of $\left(\frac{\varrho}{\varrho_0}\right)^{k-1}$, begins at the sun's center where $x=0$. For the small values of x, integration by series enables us readily to deduce from equations (6) and (7) the following approximate numerical equations:

For $k=1\cdot4$,

$$\mu=\tfrac{1}{3}x^3-\tfrac{1}{12}x^5+\tfrac{5}{336}x^7-\tfrac{125}{54432}x^9+\ \&\text{c.} \tag{8}$$

$$1-\left(\frac{1}{\varrho_0}\right)^{\cdot4}=\tfrac{1}{6}x^2-\tfrac{1}{48}x^4+\tfrac{5}{2016}x^6-\tfrac{125}{435456}x^8+\ \&\text{c.} \tag{9}$$

For $k=1\frac{2}{3}$,

$$\mu=\tfrac{1}{6}x^3-\tfrac{1}{20}x^5+\tfrac{1}{240}x^7-\tfrac{1}{3888}x^9+\ \&\text{c.} \tag{10}$$

$$1-\left(\frac{\varrho}{\varrho_0}\right)^{\frac{2}{3}}=\tfrac{1}{6}x^2-\tfrac{1}{80}x^4+\tfrac{1}{1440}x^6-\tfrac{1}{31104}x^8+\ \&\text{c.} \tag{11}$$

For larger values of x, until $\left(\frac{\varrho}{\varrho_0}\right)^{k-1}$ becomes sufficiently small as there is no need of great precision in these calculations, I have merely developed the values of μ and $\left(\frac{\varrho}{\varrho_0}\right)^{k-1}$ corresponding to $x=1\cdot1$, $x=1\cdot2$, $x=1\cdot3$, &c., by means of differences taken from the differential co-efficients at the middle of each increment of x, and for the same reason have thought it sufficient to begin with $x=1$, in equations (8) and (9) or (10) and (11). After arriving at a sufficiently small value of $\left(\frac{\varrho}{\varrho_0}\right)^{k-1}$ the calculation is finished by aid of the following approximate equations also derived by integration fron (6) and (7).

$$\mu'-\mu=\frac{k-1}{k}\,\mu'^{\frac{1}{k-1}}\,x'^{\,2-\frac{2}{k-1}}\,(x'-x)^{1+\frac{1}{k-1}}\,(1+X) \tag{12}$$

$$\left(\frac{\varrho}{\varrho_0}\right)^{k-1}=\frac{\mu'(x'-x)}{x'x}-\frac{(k-1)^2}{k(2k-1)}\mu'^{\frac{1}{k-1}}\,x'^{-\frac{2}{k-1}}\,(x'-x)^{2+\frac{1}{k-1}}$$

$$-\frac{(k-1)(2k^2-3k+2)}{k(2k-1)(3k-2)}\mu'^{\frac{1}{k-1}}\,x'^{-1-\frac{2}{k-1}}\,(x'-x)^{3+\frac{1}{k-1}} \tag{13}$$

In these equations x' and μ' are the values of x and μ corresponding to $\frac{\varrho}{\varrho_0}=0$, or the upper limit of the supposed solar atmosphere, and

$$X=-\frac{k(2k-3)}{(k-1)(2k-1)}\frac{x'-x}{x'}+\frac{k(k-2)(2k-3)}{2(k-1)^2(3k-2)}\frac{(x'-x)^2}{x'^2}+\&c.$$

$$-\frac{k-1}{2k(2k-1)}\mu'^{-1+\frac{1}{k-1}}x'^{2-\frac{2}{k-1}}(x'-x)^{1+\frac{1}{k-1}}+\&c.$$

With the values of x' and μ' determined, using r' and m' to express in like manner the corresponding values of r and m at the upper limit of the theoretical atmosphere, we find from equations (4) and (5)

$$\varrho_0=\frac{m'x'^3}{4\pi\mu'r'^3}, \tag{14}$$

$$\sigma t_1=\frac{4\pi(k-1)R^2r'^2\varrho_0^{2-k}}{kMx'^2},$$

and by equation (1), $\quad \sigma t=\dfrac{4\pi(k-1)R^2r'^2\varrho_0}{kMx'^2}\left(\dfrac{\varrho}{\varrho_0}\right)^{k-1}$, \qquad (15)

$$=\frac{k-1}{k}\frac{m'R^2x'}{\mu'Mr'}\left(\frac{\varrho}{\varrho_0}\right)^{k-1} \tag{16}$$

A glance at equation (7) will show that $\dfrac{\mu'(x'-x)}{x'x}$, equation (13), or $\dfrac{\mu'}{x'}\dfrac{r'-r}{r}$ may be taken equal to $\left(\dfrac{\varrho}{\varrho_0}\right)^{k-1}$ throughout the considerable upper part of the volume of the hypothetic gaseous body in which $1-\dfrac{\mu}{\mu'}$, or $1-\dfrac{m}{m'}$, is sufficiently small to be neglected. This substitution in the last equation gives

$$\sigma t=\frac{k-1}{k}\frac{m'R^2}{Mrr'}(r'-r),\text{ nearly,} \tag{17}$$

and also $\qquad \varrho=\left(\dfrac{\mu'}{x'}\right)^{\frac{1}{k-1}}\varrho_0\left(\dfrac{r'-r}{r}\right)^{\frac{1}{k-1}}$ nearly,

$$=\frac{1}{4\pi}\mu'^{-1+\frac{1}{k-1}}x'^{3-\frac{1}{k-1}}\frac{m'}{r'^3}\left(\frac{r'-r}{r}\right)^{\frac{1}{k-1}} \tag{18}$$

Now the mechanical equivalent of the heat in the mass ϱ of a cubic unit in volume of any perfect gas whose atmospheric subtangent is σt, is $\dfrac{1}{k-1}\varrho\cdot\sigma t$, and the mechanical equivalent of the heat that it would give out, in being cooled down under constant pressure to absolute zero, is $\dfrac{k}{k-1}\varrho\cdot\sigma t$. If the density ϱ is taken in units of the density of water, and the unit of length be the foot, this expression is multiplied by $62\frac{1}{2}$ to give for the mechanical equivalent in foot pounds

$$62\frac{1}{2}\frac{k}{k-1}\varrho\cdot\sigma t=\frac{62\frac{1}{2}}{4\pi}\mu'^{-1+\frac{1}{k-1}}x'^{3-\frac{1}{k-1}}\frac{m'^2R^2}{Mr'^4}\left(\frac{r'-r}{r}\right)^{1+\frac{1}{k-1}} \tag{19}$$

The mechanical equivalent $\dfrac{1}{k-1}\varrho\cdot\sigma t$, of the heat in the mass ϱ, viewed in the light of Clausius' mechanical theory of the gases, includes the motions of the separate atoms of each supposed compound molecule relatively to each other, as well as the motion of translation which each compound molecule makes in a straight path through free space till it impinges upon another compound molecule. If we wish to find the mechanical equivalent which would be due to this motion of translation alone, we must put $k=1\frac{2}{3}$ in the factor $\dfrac{1}{k-1}$ by which $\varrho\cdot\sigma t$ is multiplied, and this gives $\frac{3}{2}\varrho\cdot\sigma t$. To find from this the mean of the squares of the velocities of translation of the compound molecules, we divide by the mass ϱ, and, if the foot be the unit of length, multiply by $64\cdot3$, whence we have for the velocity found by taking the square root of this mean of the squares

$$8\cdot02\sqrt{\tfrac{3}{2}\sigma t}=8\cdot02\left(\frac{3}{2}\frac{k-1}{k}\frac{m'R^2x'}{\mu'Mr'}\right)^{\frac{1}{2}}\left(\frac{\varrho}{\varrho_0}\right)^{\frac{k-1}{2}} \tag{20}$$

Determination of the curve of density for $k=1\cdot4$.—Beginning with $x=1$, in equations (8) and (9), we find $\mu=\cdot2626$ and $\left(\dfrac{\varrho}{\varrho_0}\right)^{\frac{4}{10}}=\cdot8520$. Developing the values of μ and $\left(\dfrac{\varrho}{\varrho_0}\right)^{\frac{4}{10}}$ for $x=1\cdot1$, $x=1\cdot2$, &c., by means of differences we arrive at the values $\mu=2\cdot145$ and $\left(\dfrac{\varrho}{\varrho_0}\right)^{\frac{4}{10}}=\cdot1378$ when $x=4\cdot0$. Putting these values into equations (12) and (13) we find

$$x'=5\cdot355, \quad \mu'=2\cdot188.$$

If we now allow $\frac{1}{22}$d of the radius of the photosphere, or about 20,000 miles, for the height of the theoretic upper limit of the solar atmosphere above the photosphere, and if we take the mean specific gravity of the earth's mass at $5\frac{1}{2}$, and the mean specific gravity of the sun within the photosphere at $\frac{1}{4}$ that of the earth, as it is known to be, these values of x' and μ' give us in equation (14)

$$\varrho_0=28\cdot16,$$

so that the density of the sun's mass at the center would be nearly one-third greater than that of the metal platinum.

Curve of density for $k=1\frac{2}{3}$.—For this value of k the numerical coefficients in equations (8) and (9) are replaced by those in (10)

세상에서 가장 쉬운 과학 수업 별의 물리학

and (11). Otherwise, the same process employed with the value
$k_{,}=1\cdot4$, gives, starting with $x=1$, $\mu=\cdot2875$ and $\left(\dfrac{\varrho}{\varrho_0}\right)^{\frac{2}{8}}=\cdot8452$,
and developing for $x=1\cdot1$, $x=1\cdot2$, &c., brings us to $\mu=2\cdot557$
and $\left(\dfrac{\varrho}{\varrho_0}\right)^{\frac{2}{8}}=\cdot1591$, for $x=3\cdot0$, and finally gives us

$$x'=3\cdot656,\ \mu'=2\cdot741,$$

and if we now assume the same height as before for the theoretic upper limit of the sun's atmosphere, instead of $\varrho_0=28\cdot16$, we find

$$\varrho_0=7\cdot11.$$

The new curve of density is found in the same way as the first, and is presented to the eye in the diagram in comparison with it. In the upper part of both curves the scale of density is increased ten fold, and it is, in part only, evident to the eye how immensely different, for the two values of k, becomes the density in the upper parts of the sun's mass. It appears to the eye only in part because the ratio of the two densities multiplies itself rapidly in approaching the upper limit of the atmosphere.

The above was communicated in writing as here given, to the Academy at its late session.* The draft of the following, and a part of the details of its substance, have been prepared since.

Equation (20) gives in feet the square root of the mean square of velocity of translation of molecules ($8\cdot02\sqrt{\frac{3}{2}\sigma t}$). At the sun's center we find this would be 331 miles per second for the curve of density corresponding to $k=1\frac{2}{3}$, and 380 miles per second for the curve of density corresponding to $k=1\cdot4$.

In 1838 Pouillet, following the law of heat radiation given by Dulong and Petit, estimated the temperature of the radiating surface of the sun, from observations by himself of the quantity of heat it emits, at from 1461° C. to 1761° C. Herschel, from Pouillet's observations, and his own made at the Cape of Good Hope about the same time, adopts, after allowing one-third for the absorption of our atmosphere, forty feet as the thickness of ice that would be melted per minute at the sun's sur-

* I desire here to state that the formulæ which show the relation between the temperature, the pressure, the density, and the depth below the upper limit of the atmosphere, so far as they apply to the upper part of the sun's body, were independently pointed out by Prof. Peirce, in a very interesting paper which that distinguished physicist read before the Academy at the same session, and prior to the presentation of this paper. Also to recall a fact which I first learned from Prof. Peirce's mention of it to the Academy, viz. that Prof. Henry long ago threw out the idea of the atmospheric condition to which Prof. Thomson has more recently given the term convective equilibrium, viz., such that any portion of the air, on being conveyed into any new layer above or below, would find itself reduced, by its expansion or compression, to the temperature of the new layer.

face. The temperature of the radiating surface calculated from this datum by the formula of Dulong and Petit, and with the co-efficient of radiation found by Prof. W. Hopkins for sandstone, the smallest co-efficient he found, is 1550° C. or 2820° Fah. But then the solar radiation is many thousands of times greater than the greatest in Dulong and Petit's experiments, so that these calculations of the temperature of the sun's photosphere have little weight notwithstanding the simplicity and accuracy with which the formula represents the experiments from which it was derived. Nothing authorizes us to accept the formula as more than an empirical one. It seems desirable that experiments similar to those of Dulong and Petit should be made on the rate of cooling of intensely heated bodies, such as balls of platinum not too large. By placing the heated ball in the center of a hollow spherical jacket of water, either flowing or in an unchanged mass, the quantities of heat radiated in successive equal spaces of time will be determined, and the corresponding differences of temperature in the heated ball can at least be estimated with whatever probability we may rely on our knowledge of the specific heat of its material. At present the best means we have of forming any judgment of the probable temperature of the source of the sun's radiation, is perhaps to be found in a comparison between the effects of the hydro-oxygen blowpipe, and the recorded effects of Parker's great burning lens. I am not aware that this method has before been resorted to.

If the angle of aperture at the focus of a burning lens, or combination of lenses, be called $2a$, the radiation received by a small flat surface at the focus will be $\sin^2 a$, if a unit be taken to represent the radiation the same small flat surface would receive just at the sun's surface. Parker's lens, with the small lens added, had, at the focus so formed, an angle of aperture of about 47°. A small flat surface at its focus would therefore receive about one-sixth the radiation that it would just at the sun, making no allowance for absorption by the atmospheres of the earth and sun and rays lost in transmission through the lenses. Pouillet, from the experiments already alluded to made by himself, found his atmosphere in fine weather transmitted, of the sun's heat rays, about the fraction $\frac{3}{4}$ raised to a power whose exponent is the secant of the sun's zenith distance. This, of course, leaves out of view the heat rays of low intensity which are totally absorbed by the atmosphere. He also concluded from comparison with other experiments of his own with a moderately large burning glass, that that glass transmitted $\frac{7}{8}$ of the heat rays incident on it. If we assume the same fraction for each of the two lenses of Parker's com-

세상에서 가장 쉬운 과학 수업 별의 물리학

bination, and assume further that the sun's zenith distance did not exceed 48° in the experiments made with it, we find for the fractional multiplier expressing the part of the sun's heat radiation which arrived at the focus unintercepted, $(\frac{4}{5})^{1\cdot 1\cdot 2}(\frac{7}{8})^2 = \cdot 55$. Hence the radiation actually received by a small flat surface at the focus was ·09, or about one-eleventh, of what it would receive just at the sun. The heat so received by any body so placed in the focus, must, after the body has acquired its highest temperature, be emitted from it at the same rate. The heat so emitted will consist: first, of heat radiated into that part of space toward which the radiating surface of the body looks; secondly, of heat carried of by convection of the air; thirdly, of heat conducted away by the body supporting the body subjected to experiment; fourthly, of heat rays, if any, reflected, and not absorbed, by the body subjected to experiment. Assuming it as a reasonable conjecture that full half of all this* consists of heat *radiated* into the single *hemisphere* looking upon a flat surface, we may conclude that the body, at its highest acquired temperature, radiated not less than $\frac{1}{20}$th as much heat as is radiated by an equal extent of surface of the sun's photosphere, over and above such part of that radiation as may be intercepted by the sun's atmosphere, and such rays of low intensity as are *totally* absorbed by our own atmosphere, the whole of which apparently cannot be great. No allowance seems necessary for the chromatic and spherical dispersion of the lenses, since the diameter of the focus is stated at half an inch, while the true diameter of the sun's image would be not less than one-third of an inch.

Now we are not without the means of forming a probable approximate estimate of this temperature at which the radiation becomes $\frac{1}{20}$th, more or less, of that of the sun's photosphere. We are told that in the focus of Parker's compound lens 10 grains of very pure lime ("white rhomboidal spar") were melted in 60 seconds. We may presume that in that length of time the temperature of the lime, after parting with its carbonic acid, made a near approximation to the maximum at which it would be stationary, a presumption confirmed by the period of 75 seconds said to have been occupied in the fusion of 10 grains of carnelian, and by the considerable period of 45 seconds for the fusion of a topaz of only 3 grains, and 25 seconds for an oriental emerald of but 2 grains, and in fact sufficiently

* As to the heat carried off by convection of the air, if its quantity be calculated by the formula given by Dulong and Petit for that purpose, it comes out utterly insignificant in comparison with the heat received from the burning glass. The conjectural allowance of $\frac{4}{5}$ths in all, of this, is likely, therefore, to be much too large. Not much reliance, indeed, can be placed upon the formula here mentioned, at such a temperature as 4000° Fah., yet, as by it the convection is taken proportional to the 1·233 power of the difference of temperature, it seems unlikely that it gives a quantity very many fold less than the truth.

proved, it would seem, by observing that the heat we have estimated to fall at the focus, upon a flat surface, would suffice, if retained, to raise the temperature of a quarter of an inch thick of lime 4000° Fah. in 5 seconds. If, then, we may take the temperature *maintained* at the focus of Parker's lens to have been at the melting point of lime, we may conclude that it is also not far from the temperature given by the hydro-oxygen blowpipe. Dr. Hare, who was the first inventor of this instrument, and the discoverer of its great power, melted down, by its means, in partial fusion, a very small stick of lime cut on a lump of that material, which we understand to have been a very pure specimen. Burning glass and blowpipe seem each to have been near the limit of its power in this apparently common effect. But Deville found the temperature produced by the combination of hydrogen and oxygen under the atmospheric pressure to be 2500° Cent. As the lime in the heated blast would radiate rapidly, its temperature must have been lower than that of combined hydrogen and oxygen, and I have called it 2220° Cent. or 4000° Fah.

The formula of Dulong and Petit, with the co-efficient found by Hopkins, as already mentioned, gives for the quantity of heat radiated in one minute by a square foot of surface of a body whose temperature is $\theta+t$ centigrade, into a chamber whose temperature is θ centigrade, when expressed with the unit employed by Hopkins,

$$8\cdot377\,(1\cdot0077)^{\theta}\,[(1\cdot0077)^{t}-1].$$

It will be convenient, and, in the discussion of the high temperatures with which we are concerned, will involve no sensible error, to use the hypothesis that the space around the radiating body is at the temperature of 0° C. and the formula for the radiation then becomes,

$$8\cdot377\,[(1\cdot0077)^{t}-1]. \qquad (21)$$

The unit used by Hopkins, in the formula here given, is the quantity of heat that will raise the temperature of 1000 grains of water 1° centigrade. Expressed by the same unit, the quantity adopted by Sir J. Herschel as the amount of the sun's radiation, viz. that which would melt 40 feet thick of ice in a minute (at the sun's surface), is 1,280,000. The $\frac{1}{20}$th of this, or 64,000, expresses, therefore, the quantity which we have estimated the lime under Parker's lens to have radiated, per square foot of its surface, at its estimated temperature of 4000° Fah. If now we calculate its temperature by the above formula, from the estimated radiation, the result is 1166° Cent. or 2130° Fah. This is manifestly much below the real temperature, and so far below that there can be no doubt the formula of Dulong and Petit has failed at the melting point of lime. If instead of the co-efficient 8·377 we had used the larger co-efficient 12·808 which Hopkins gives for unpolished limestone, the formula would have been reduced only 53° Cent. It best suits the direction of our inquiry to use the smallest co-efficient which Hopkins' experiments gave, since we are seeking the *highest* temperature which can be plausibly deduced from the sun's radiation. For ease of expression, the curve which we will imagine for representing the actual relation of radiation to temperature, the horizontal ordinate standing for the temperature and the vertical ordinate for the radiation corresponding thereto, may be called the curve of radiation. The course of this curve from the freezing point of water to a point somewhat below the boiling point of mercury is correctly marked out to us by the formula. Beyond that we have but the rough approximation which we can get by means of the above comparison, to the single point of the curve where the radiation is $\frac{1}{20}$th that of the sun's photosphere. The attempt, from these data, to extend the curve till it reaches the full radiation of that photosphere, must be mainly conjectural. As a basis for the most plausible conjecture I am able to make let us assume : first, that the upward concavity of the curve of radiation, which increases very rapidly with the temperature as far as the curve follows the formula of Dulong and Petit, is at no temperature greater than that formula would give it at the same temperature ; secondly, that the curve of radiation is nowhere convex upward. If, then, we set out from these two conjectural assumptions—of the degree of probability of which each one must form his own impression—the greatest temperature the sun's photosphere could have consistently with the radiation of 64,000 at the temperature of 4000° Fah., is found by drawing through the point representing that radiation and that temperature a straight line tangent to the curve of the formula. The line so drawn would cross the real curve of radiation in a greater or less angle at the radiation of 64,000 and temperature of 4000° Fah., and at higher temperatures would fall more or less *below* that curve, and its intersection with the sun's radiation of 1,280,000 would be at a temperature greater than that of the curve, that is to say, greater than the temperature of the sun's photosphere. This greater temperature is 55,450° Fah.

A different train of conjecture led me at first to assume a temperature of 54,000° Fah., and this last number I will here retain since it has been already used as the basis of some of the calculations we now proceed to give. It must be here recollected that we are discussing the question of *clouds* of *solid* or at least *fluid* particles floating in non-radiant gas, and constituting the sun's photosphere. If the amount of *radiation*

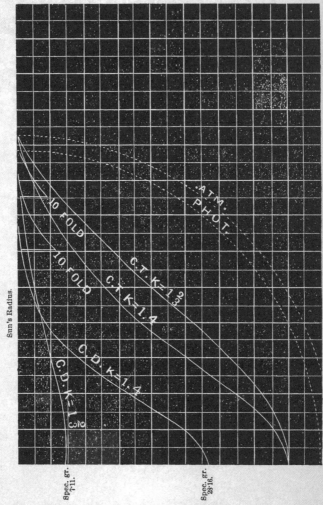

Explanation.—ATM., Assumed theoretic upper limit of atmosphere; PHOT., Photosphere; C.T.K.=$1\frac{2}{3}$, Arbitrary Curve of temperature for $k=1\frac{2}{3}$; C.T.K.=1·4, Arbitrary Curve of temperature for $k=1·4$; C.D.K.=1·4, Absolute Curve of density for $k=1·4$; C.D.K.=$1\frac{2}{3}$, Absolute density for $k=1\frac{2}{3}$.

세상에서 가장 쉬운 과학 수업 별의 물리학

would lead us to limit the temperature of such clouds of solids or fluids, so also it seems difficult to credit the *existence* in the solid or fluid form, at a higher temperature than 54,000° Fah. of any substance that we know of.

If then we suppose a temperature of 54,000° Fah., what would be the density of that layer of the hypothetic gaseous body which has that temperature, and what length of time would be required, at the observed rate of solar radiation, for the emission of all the heat that a foot thick of that layer would give out in cooling down under pressure to absolute zero? The latter question depends on the mechanical equivalent of this heat for a cubic foot of the layer of gas, and the two questions, together with that of the depth at which the layer would be situated below the theoretic upper limit of the atmosphere, are answered by equations (17), (18), and (19), provided we knew the value of k and the value of σ in the body of gas. The less the atomic weight of the gas the greater the value of σ, and the greater the density of the layer of 54000° Fah. and the greater the quantity of heat which a cubic foot of it would give out in cooling down. I therefore base the first calculation on hydrogen as it is known to us. The value of σ is in that case about 800 feet, and the value of k about 1·4, nearly the same as in common air. These values would give for the layer of 54000° Fah. a specific gravity about ·00000095 that of water, or about one 90th that of hydrogen gas at common temperature and pressure, and the mechanical equivalent of the heat that a cubic foot of the layer would give out in cooling down under pressure to absolute zero would be only about 9000 foot pounds, whereas the mechanical equivalent of the heat radiated by one square foot of the sun's surface in one minute is about 254,000,000 foot pounds. The heat emitted each minute would, therefore, be fully half of all that a layer ten miles thick would give out in cooling down to zero, and a circulation that would dispose of volumes of cooled atmosphere at such a rate seems inconceivable.

It may possibly appear to some minds that the difficulty presented by this aspect of the case will vanish if we suppose the photosphere, or its cloudy particles, to be maintained by radiation at a temperature to almost any extent lower than that of convective equilibrium. This would enable us to place the theater of operations in a lower and denser layer of atmosphere, but the supposition seems to me difficult to realize unless, as the hot gases rise from beneath, precipitation could commence at a temperature many times higher than the 54000° Fah. which we have estimated for the upper visible surface of the clouds, and this, as before intimated, seems to me itself extremely improbable.

I may mention here that my friend Dr. Craig, in an unpub-lished paper, following the hint thrown out by Frankland, is disposed to favor the idea that the sun's radiation may be the radiation of hot gases instead of clouds. At present I shall offer no opinion on that point one way or the other, but will only state it as my impression that if the theory of precipitated clouds, as above presented, is the true one, something quite unlike our present experimental knowledge, or at least much beyond it, is needed to make it intelligible.

The first hypothesis which offers itself in an attempt to make the theory rational is suggested by one point in Clausius' theory of the constitution of the gases, already alluded to. In forming his theory Clausius found that the known specific heats of the gases are all much too great for free simple atoms impinging on one another, and he therefore introduced the hypothesis of compound molecules, each compound molecule being a system of atoms oscillating among each other under forces of mutual attraction. Now if this were accepted as the actual constitution of the gases it is of course easy enough to conceive that in the fierce collisions of these compound mole-cules with each other at the temperatures supposed to exist in the sun's body, their component atoms might be torn asunder, and might thenceforth move as free simple molecules. In this case, still retaining the hypothesis of Clausius' theory, that the average length of the path described by each between collisions is large compared with the diameter of the sphere of effective attraction or repulsion of atom for atom, the value of k would reach its maximum of $1\frac{2}{3}$. Experiment has not shown us any gas in this condition, and for the present it is hypothetical. Even in hydrogen the value of k does not materially, if any, exceed the value of 1.4 which it has in air. But if it were found that the hydrogen molecule is compound, and that in the body of the sun the heat splits this molecule into two equal simple atoms, and in fact that all the matter in the sun's body is split into simple free atoms equally as small, then, while the value of k would be $1\frac{2}{3}$, the value of σ would be about 1600 feet. If with these values we repeat the calculation of the density of the layer of 54000° Fah. we find its specific gravity to be 0.000363 of that of water, or 4.35 times that of hydrogen gas at common temperature and pressure and in its known con-dition, or 8.7 times that which the hydrogen in the hypothetic condition would have if it retained that condition at common temperature and pressure. We find also that the mechanical equivalent of all the heat that a cubic foot of the layer would give out in cooling down, under pressure, to zero, would be no less than 13,500,000 foot pounds. Instead, therefore, of a layer ten miles thick, it would now require only a thickness of 38 feet

세상에서 가장 쉬운 과학 수업 별의 물리학

to give out, in cooling down to zero, twice the heat emitted by the sun in one minute. It will be seen, (equations (17) and (19)), that this thickness, retaining the constant value $k=1\frac{2}{3}$, would diminish with the $2\frac{1}{2}$ power of the masses of the atoms into which the sun's body is hypothetically resolved (the reciprocal of the value of σ), and I leave each to form his own impression how far this view leads towards verisimilitude.

It is important to add that the depth of the layer of 54000° Fah. below the theoretic upper limit of atmosphere, when calculated with value $k=1\cdot4$, $\sigma=800$ feet, comes out only 1107 miles, and with the values $k=1\frac{2}{3}$ and $\sigma=1600$ feet only 1581 miles. This calculation of the depth, unlike the other results above, may be said to be independent of the question of the constitution of the sun's interior mass. It is alike difficult, on any plausible hypothesis, to reconcile a temperature no higher than 54000° Fah. with any perceptible atmosphere extending many thousand miles above, and yet no less an authority than Prof. Peirce has assigned a hundred thousand miles as the height of the solar atmosphere above the photosphere, at the same time, however, pointing out the enormous temperature which, under convective equilibrium, this would imply at the level of the photosphere. But all are not yet agreed that the appearances seen at such distances from the sun are proof of the existence of a true atmosphere there. It will be seen that the numbers I give above were obtained from a first hypothesis of an atmospheric limit 20,000 miles above the photosphere, but for the purpose of this paper it is of no consequence to repeat the calculation from a different limit.

It is, I believe, recognized on theoretical grounds that in an atmosphere containing a mixture of gases of unequal density the lighter gases might be expected to diffuse in greater proportion into the higher parts of the atmosphere and the heavier gases into the lower parts. But perhaps the supposed circulation which the emission of heat maintains within the photosphere must renew mixture at a rate sufficient to mask the rate which theory would assign for diffusion. I have not attempted a theoretic comparison between these two tendencies. It will suffice here to repeat that the above numerical results, so far as they may be thought to give countenance to the theory in its mechanical aspect, require that the entire inner mass of the sun shall have, at a mean, (in the supposed state of dissociation), the very small atomic weight specified. We may notice in this connection the uniform proportion of oxygen and nitrogen gases in our atmosphere at the height of four miles or more at which the analysis has been made. Without having gone into a critical examination of the question, I suppose that at that height the proportion of oxygen which the theory of diffusive equilibrium would assign is notably diminished, and that it would be found that the circulation of the air is sufficiently active to mask the theoretic rate of diffusion.

The second hypothesis which might offer itself in an attempt to make the theory rational, but which a very little reflection is, I think, sufficient to set aside, is that which would modify Clausius' theory of the gases by assuming that in the sun's body the average length of the excursion made by each molecule between two consecutive collisions, becomes very short compared with the radius of the sphere of repulsion of molecule for molecule, and with the average distance of their centers at nearest approach. This way of harmonizing the actual volume of the sun with such a temperature as 54000° Fah. in the photosphere, and with the smallest density which we can credit in the photosphere, would involve the consequence that the existing density of almost the entire mass of the sun is very nearly uniform and at its maximum possible, or at all events that any further sensible amount of collapse must be productive of but a very small amount, comparatively, of renewed supplies of heat, for the obvious reason that this hyphothesis carries with it almost the entire neutralization of the force of gravity by the forces of molecular repulsion. In like manner it involves the consequence that in any such small contraction of the photosphere as can have taken place within the history of total eclipses, it is but a very small fraction of the sun's mass, near its surface, that can have taken part in the collapse to any thing like a proportionate extent. Hence it also extremely restricts the period during which we could suppose the sun to have existed under anything like its present visible magnitude in the past, consistently with the production in the way supposed of the supplies of heat it has been sending out. Another thing involved in this second hypothesis is the fact which Prof. Peirce has pointed out to the Academy, viz.: that the existing molecular repulsion in the sun's body would immensely exceed such as would be indicated by the modulus of elasticity of any form of matter known to us.

In conclusion, I do not mean to say that there is any *invincible logical exclusion* of any law of the action of gases different from what is specified or alluded to above. I only mean that, so far as I can see, any theory of heat which is based *simply* and *solely* upon *molecular attraction* and *repulsion* dependent on molecular distance alone, cannot in its application to the sun, escape from the conditions indicated in this paper. It is certainly not absurd to imagine heat to be an agent of some kind so constituted that it cannot be thus represented by the sole conditions of motion and of molecular attraction and repulsion, but yet so constituted that in its effects upon matter it follows the conditions of mechanical equivalency as defined by Joule. In fact, such exceptional cases as the expansion of water in freezing seem to favor such a view, though the range of that phenomenon is very limited. One way of forming a mechanical representation of such a constitution would be by associating molecular motion with the mechanical powers, either with or without molecular attraction or repulsion; the manner in which the imagined mechanical power (or link) attaches itself to the molecules which it connects—so as to make their motion determine their mutual approach or recession or change of relative direction—being dependent on the existing motions and other conditions in such a way as to produce the observed phenomena. The possibility of such a mechanical *representation* is sufficient to show that such a supposed *constitution* is not *logically excluded*, but to accept such a mechanical representation as a physical fact is quite another matter, and, as it seems to me, a very difficult one. Of course this difficulty does not present itself when we suppose that heat is not motion.

논문 웹페이지

The Internal Constitution of the Stars.*
By Prof. A. S. Eddington, M.A., M.Sc., F.R.S.

September 2, 1920

L AST year at Bournemouth we listened to a proposal from the President of the Association to bore a hole in the crust of the earth and discover the conditions deep down below the surface. This proposal may remind us that the most secret places of Nature are, perhaps, not 10 to the nth miles above our heads, but 10 miles below our feet. In the last five years the outward march of astronomical discovery has been rapid, and the most remote worlds are now scarcely safe from its inquisition. By the work of H. Shapley the globular clusters, which are found to be at distances scarcely dreamt of hitherto, have been explored, and our knowledge of them is in some respects more complete than that of the local aggregation of stars which includes the sun. Distance lends not enchantment, but precision, to the view. Moreover, theoretical researches of Einstein and Weyl make it probable that the space which remains beyond is not illimitable; not merely the material universe, but also space itself, is perhaps finite; and the explorer must one day stay his conquering march for lack of fresh realms to invade. But to-day let us turn our thoughts inwards to that other region of mystery—a region cut off by more substantial barriers, for, contrary to many anticipations, even the discovery of the fourth dimension has not enabled us to get at the inside of a body. Science has material and non-material appliances to bore into the interior, and I have chosen to devote this address to what may be described as analytical boring devices —*absit omen!*

The analytical appliance is delicate at present, and, I fear, would make little headway against the solid crust of the earth. Instead of letting it blunt itself against the rocks, let us look round for something easier to penetrate. The sun? Well, perhaps. Many have struggled to penetrate the mystery of the interior of the sun; but the difficulties are great, for its sub-

세상에서 가장 쉬운 과학 수업 별의 물리학

stance is denser than water. It may not be quite so bad as Biron makes out in " Love's Labour's Lost " :

The heaven s glorious sun
That will not be deep-search'd with saucy looks :
Small have continual plodders ever won
Save base authority from others' books.

But it is far better if we can deal with matter in that state known as a perfect gas, which charms away difficulties as by magic. Where shall it be found?

A few years ago we should have been puzzled to say where, except perhaps in certain nebulæ; but now it is known that abundant material of this kind awaits investigation. Stars in a truly gaseous state exist in great numbers, although at first sight they are scarcely to be discriminated from dense stars like our sun. Not only so, but the gaseous stars are the most powerful light-givers, so that they force themselves on our attention. Many of the familiar stars are of this kind —Aldebaran, Canopus, Arcturus, Antares; and it would be safe to say that three-quarters of the naked-eye stars are in this diffuse state. This remarkable condition has been made known through the researches of H. N. Russell (NATURE, vol. xciii., pp. 227, 252, 281) and E. Hertzsprung; the way in which their conclusions, which ran counter to the prevailing thought of the time, have been substantiated on all sides by over-whelming evidence is the outstanding feature of recent progress in stellar astronomy.

The diffuse gaseous stars are called *giants,* and

*Opening address of the president of Section A (Mathematical and Physical Science) delivered at the Cardiff meeting of the British Association on August 24.

the dense stars *dwarfs.* During the life of a star there is presumably a gradual increase of density through contraction, so that these terms distinguish the earlier and later stages of stellar history. It appears that a star begins its effective life as a giant of comparatively low temperature—a red or M-type star. As this diffuse mass of gas contracts its temperature must rise, a conclusion long ago pointed out by Homer Lane. The rise continues until the star becomes too dense, and ceases to behave as a perfect gas. A maximum temperature is attained, depending on the mass, after which the star, which has now become a dwarf, cools and further contracts. Thus each temperature-level is passed through twice, once in an ascending and once in a descending stage—once as a giant, once as a dwarf. Temperature plays so predominant a part in the usual spectral classification that the ascending and descending stars were not originally discriminated, and the customary classification led to some perplexities. The separation of the two series was discovered through their great difference in luminosity, particularly striking in the case of the red and yellow stars, where the two stages fall widely apart in the star's history. The bloated giant has a far larger surface than the compact dwarf, and gives correspondingly greater light. The distinction was also revealed by direct determinations of stellar densities, which are possible in the case of eclipsing variables like Algol. Finally, Adams and Kohlschütter have set the seal on this discussion by showing that there are actual spectral differences between the ascending and descending stars at the same temperature-level, which are conspicuous enough when they are looked for.

Perhaps we should not too hastily assume that the direction of evolution is necessarily in the order of increasing density, in view of our ignorance of the origin of a star's heat, to which I must allude later. But, at any rate, it is a great advance to have disentangled what is the true order of continuous increase of density, which was hidden by superficial resemblances.

The giant stars, representing the first half of a star's life, are taken as material for our first boring experiment. Probably, measured in time, this stage corresponds to much less than half the life, for here it is the ascent which is easy and the way down is long and slow. Let us try to picture the conditions inside a giant star. We need not dwell on the vast dimensions—a mass like that of the sun, but swollen to much greater volume on account of the low density, often below that of our own atmosphere. It is the star as a storehouse of heat which especially engages our attention. In the hot bodies familiar to us the heat consists in the energy of motion of the ultimate particles, flying at great speeds hither and thither. So, too, in the stars a great store of heat exists in this form; but a new feature arises. A large proportion, sometimes more than half the total heat, consists of imprisoned radiant energy—aether-waves travelling in all directions trying to break through the material which encages them. The star is like a sieve, which can retain them only temporarily; they are turned aside, scattered, absorbed for a moment, and flung out again in a new direction. An element of energy may thread the maze for hundreds of years before it attains the freedom of outer space. Nevertheless, the sieve leaks, and a steady stream permeates outwards, supplying the light and heat which the star radiates all round.

That some æthereal heat as well as material heat

세상에서 가장 쉬운 과학 수업 별의 물리학

exists in any hot body would naturally be admitted; but the point on which we have here to lay stress is that in the stars, particularly in the giant stars, the æthereal portion rises to an importance which quite transcends our ordinary experience, so that we are confronted with a new type of problem. In a red-hot mass of iron the æthereal energy constitutes less than a billionth part of the whole; but in the tussle between matter and æther the æther gains a larger and larger proportion of the energy as the temperature rises. This change in proportion is rapid, the æthereal energy increasing rigorously as the fourth power of the temperature, and the material energy roughly as the first power. But even at the temperature of some millions of degrees attained inside the stars there would still remain a great disproportion; and it is the low density of material, and accordingly the reduced material energy per unit volume in the giant stars, which wipes out the last few powers of 10. In all the giant stars known to us, widely as they differ from one another, the conditions are just reached at which these two varieties of heat-energy have attained a rough equality; at any rate, one cannot be neglected compared with the other. Theoretically there could be conditions in which the disproportion was reversed and the æthereal far outweighed the material energy; but we do not find them in the stars. It is as though the stars had been measured out—that their sizes had been determined—with a view to this balance of power; and one cannot refrain from attributing to this condition a deep significance in the evolution of the cosmos into separate stars.

To recapitulate. We are acquainted with heat in two forms—the energy of motion of material atoms and the energy of æther waves. In familiar hot bodies the second form exists only in insignificant quantities. In the giant stars the two forms are present in more or less equal proportions. That is the new feature of the problem.

On account of this new aspect of the problem the first attempts to penetrate the interior of a star are now seen to need correction. In saying this we do not depreciate the great importance of the early researches of Lane, Ritter, Emden, and others, which not only pointed the way for us to follow, but also achieved conclusions of permanent value. One of the first questions they had to consider was by what means the heat radiated into space was brought up to the surface from the low level where it was stored. They imagined a bodily transfer of the hot material to the surface by currents of convection, as in our own atmosphere.

But actually the problem is, not how the heat can be brought to the surface, but how the heat in the interior can be held back sufficiently—how it can be barred in and the leakage reduced to the comparatively small radiation emitted by the stars. Smaller bodies have to manufacture the radiant heat which they emit, living from hand to mouth; the giant stars merely leak radiant heat from their store. I have put that much too crudely; but perhaps it suggests the general idea.

The recognition of æthereal energy necessitates a twofold modification in the calculations. In the first place, it abolishes the supposed convection currents; and the type of equilibrium is that known as radiative instead of convective. This change was first suggested by R. A. Sampson so long ago as 1894. The detailed theory of radiative equilibrium is particularly associated with K. Schwarzschild, who applied it to the sun's atmosphere. It is perhaps still uncertain whether it holds strictly for the atmospheric layers, but the arguments for its validity in the interior of a star are far more cogent. Secondly, the outflowing stream of æthereal energy is powerful enough to exert a *direct mechanical effect* on the equilibrium of a star. It is as though a strong wind were rushing outwards. In fact, we may fairly say that the stream of radiant energy *is* a wind; for though æther waves are not usually classed as material, they have the chief mechanical properties of matter, viz. mass and momentum. This wind distends the star and relieves the pressure on the inner parts. The pressure on the gas in the interior is not the full weight of the superincumbent columns, because that weight is partially borne by the force of the escaping æther waves beating their way out. This force of radiation-pressure, as it is called, makes an important difference in the formulation of the conditions for equilibrium of a star.

Having revised the theoretical investigations in accordance with these considerations (*Astrophysical Journal,* vol. xlviii., p. 205), we are in a position to deduce some definite numerical results. On the observational side we have fairly satisfactory knowledge of the masses and densities of the stars and of the total radiation emitted by them; this knowledge is partly individual and partly statistical. The theoretical analysis connects these observational data on the one hand with the physical properties of the material inside the star on the other. We can thus find certain information as to the inner material, as though we had actually bored a hole. So far as can be judged, there are only two physical properties of the material which can concern us—always provided that it is sufficiently

238 　　　　　　　　　　　　세상에서 가장 쉬운 과학 수업 별의 물리학

rarefied to behave as a perfect gas—viz. the average molecular weight and the transparency or permeability to radiant energy. In connecting these two unknowns with the quantities given directly by astronomical observation we depend entirely on the well-tried principles of conservation of momentum and the second law of thermodynamics. If any element of speculation remains in this method of investigation, I think it is no more than is inseparable from every kind of theoretical advance.

We have, then, on one side the mass, density, and output of heat, quantities as to which we have observational knowledge; on the other side, molecular weight and transparency, quantities which we want to discover.

To find the transparency of stellar material to the radiation traversing it is of particular interest, because it links on this astronomical inquiry to physical investigations now being carried on in the laboratory, and to some extent it extends those investigations to conditions unattainable on the earth. At high temperatures the æther waves are mainly of very short wave-length, and in the stars we are dealing mainly with radiation of wave-length 3 to 30 Ångström units, which might be described as very soft X-rays. It is interesting, therefore, to compare the results with the absorption of the harder X-rays dealt with by physicists. To obtain an exact measure of this absorption in the stars we have to assume a value of the molecular weight; but fortunately the extreme range possible for the molecular weight gives fairly narrow limits for the absorption. The average weight of the ultimate independent particles in a star is probably rather low, because in the conditions prevailing there the atoms would be strongly ionised; that is to say, many of the outer electrons of the system of the atom would be broken off; and as each of these free electrons counts as an independent molecule for present purposes, this brings down the average weight. In the extreme case (probably not reached in a star) when the whole of the electrons outside the nucleus are detached the average weight comes down to about 2, *whatever the material,* because the number of electrons is about half the atomic weight for all the elements (except hydrogen). We may, then, safely take 2 as the extreme lower limit. For an upper limit we might perhaps take 200; but to avoid controversy we shall be generous and merely assume that the molecular weight is not greater than—infinity. Here is the result :—

> For molecular weight 2, mass-coefficient of absorption = 10 C.G.S. units.
> For molecular weight ∞, mass-coefficient of absorption = 130 C.G.S. units.

The true value, then, must be between 10 and 130. Partly from thermodynamical considerations, and partly from further comparisons of astronomical observation with theory, the most likely value seems to be about 35 C.G.S. units, corresponding to molecular weight 3.5.

Now this is of the same order of magnitude as the absorption of X-rays measured in the laboratory. I think the result is in itself of some interest, that in such widely different investigations we should approach the same kind of value of the opacity of matter to radiation. The penetrating power of the radiation in the star is much like that of X-rays; more than half is absorbed in a path of 20 cm. at atmospheric density. Incidentally, this very high opacity explains why a star is so nearly heat-tight, and can store vast supplies of heat with comparatively little leakage.

So far this agrees with what might have been anticipated; but there is another conclusion which physicists would probably not have foreseen. The giant series comprises stars differing widely in their densities and temperatures, those at one end of the series being on the average about ten times hotter throughout than those at the other end. By the present investigation we can compare directly the opacity of the hottest stars with that of the coolest. The rather surprising result emerges that the opacity is the same for all; at any rate, there is no difference large enough for us to detect. There seems no room for doubt that at these high temperatures the absorption-coefficient is approaching a limiting value, so that over a wide range it remains practically constant. With regard to this constancy, it is to be noted that the temperature is concerned twice over : it determines the character and wave-length of the radiation to be absorbed, as well as the physical condition of the material which is absorbing. From the experimental knowledge of X-rays we should have expected the absorption to vary very rapidly with the wave-length, and therefore with the temperature. It is surprising, therefore, to find a nearly constant value

The result becomes a little less mysterious when we consider more closely the nature of absorption. Absorption is not a continuous process, and after an atom has absorbed its quantum it is put out of action for a time until it can recover its original state. We know very little of what determines the rate of recovery of the atom, but it seems clear that there is a limit to the amount of absorption that can be performed by an atom in a given time. When that limit is reached no increase in the intensity of the incident radiation

세상에서 가장 쉬운 과학 수업 별의 물리학

will lead to any more absorption. There is, in fact, a saturation effect. In the laboratory experiments the radiation used is extremely weak; the atom is practically never caught unprepared, and the absorption is proportional to the incident radiation. But in the stars the radiation is very intense and the saturation effect comes in.

Even granting that the problem of absorption in the stars involves this saturation effect, which does not affect laboratory experiments, it is not very easy to understand theoretically how the various conditions combine to give a constant absorption-coefficient independent of temperature and wave-length. But the astronomical results seem conclusive. Perhaps the most hopeful suggestion is one made to me a few years ago by C. G. Barkla. He suggested that the opacity of the stars may depend mainly on *scattering* rather than on true atomic absorption. In that case the constancy has a simple explanation, for it is known that the coefficient of scattering (unlike true absorption) approaches a definite constant value for radiation of short wave-length. The value, moreover, is independent of the material. Further, scattering is a continuous process, and there is no likelihood of any saturation effect; thus for very intense streams of radiation its value is maintained, whilst the true absorption may sink to comparative insignificance. The difficulty in this suggestion is a numerical discrepancy between the known theoretical scattering and the values already given as deduced from the stars. The theoretical coefficient is only 0·2 compared with the observed value 10 to 130. Barkla further pointed out that the waves here concerned are not short enough to give the ideal coefficient; they would be scattered more powerfully, because under their influence the electrons in any atom would all vibrate in the same phase instead of in haphazard phases. This might help to bridge the gap, but not sufficiently. It must be remembered that many of the electrons have broken loose from the atom and do not contribute to the increase.[1] Making all allowances for uncertainties in the data, it seems clear that the astronomical opacity is definitely higher than the theoretical scattering. Very recently, however, a new possibility has opened up which may possibly effect a reconciliation. Later in the address I shall refer to it again.

Astronomers must watch with deep interest the investigations of these short waves, which are being pursued in the laboratory, as well as the study of the conditions of ionisation by both experimental and theoretical physics, and I am glad of this opportunity

of bringing before those who deal with these problems
the astronomical bearing of their work.

I can allude only very briefly to the purely astronomical results which follow from this investigation
(Monthly Notices, vol. lxxvii., pp. 16, 596; vol. lxxix.,
p. 2); it is here that the best opportunity occurs for
checking the theory by comparison with observation,
and for finding out in what respects it may be defi-
cient. Unfortunately, the observational data are
generally not very precise, and the test is not so stringent as we could wish. It turns out that (the opacity
being constant) the total radiation of a giant star
should be a function of its mass only, independent of
its temperature or state of diffuseness. The total
radiation (which is measured roughly by the luminosity) of any one star thus remains constant during
the whole giant stage of its history. This agrees with
the fundamental feature, pointed out by Russell in
introducing the giant and dwarf hypothesis, that giant
stars of every spectral type have nearly the same
luminosity. From the range of luminosity of these
stars it is now possible to find their range of mass.
The masses are remarkably alike—a fact already suggested by work on double stars. Limits of mass in the
ratio 3 : 1 would cover the great majority of the giant
stars. Somewhat tentatively we are able to extend
the investigation to dwarf stars, taking account of the
deviations of dense gas from the ideal laws and using
our own sun to supply a determination of the unknown
constant involved. We can calculate the maximum
temperature reached by different masses; for example,
a star must have at least $\frac{1}{2}$ the mass of the sun in
order to reach the lowest spectral type, M; and in
order to reach the hottest type, B, it must be at least
$2\frac{1}{2}$ times as massive as the sun. Happily for the

---

[1] *E.g.* for iron non-ionised the theoretical scattering is 5·2, against an
astronomical value *120*. If 16 electrons (2 rings) are broken off, the
theoretical coefficient is 0·9, against an astronomical value 35. For different
assumptions as to ionisation the values chase one another, but cannot be
brought within reasonable range.

---

theory, no star has yet been found with a mass less
than $\frac{1}{2}$ of the sun's; and it is a well-known fact, discovered from the study of spectroscopic binaries,
that the masses of the B stars are large compared
with those of other types. Again, it is possible to
calculate the difference of brightness of the giant and
dwarf stars of type M, *i.e.* at the beginning and end
of their career; the result agrees closely with the
observed difference. In the case of a class of variable
stars in which the light changes seem to depend on a
mechanical pulsation of the star, the knowledge we
have obtained of the internal conditions enables us to

predict the period of pulsation within narrow limits. For example, for δ Cephei, the best-known star of this kind, the theoretical period is between four and ten days, and the actual period is $5\frac{1}{3}$ days. Corresponding agreement is found in all the other cases tested.

Our observational knowledge of the things here discussed is chiefly of a rather vague kind, and we can scarcely claim more than a general agreement of theory and observation. What we have been able to do in the way of tests is to offer the theory a considerable number of opportunities to "make a fool of itself," and so far it has not fallen into our traps. When the theory tells us that a star having the mass of the sun will at one stage in its career reach a maximum effective temperature of 9000° (the sun's effective temperature being 6000°) we cannot do much in the way of checking it; but an erroneous theory might well have said that the maximum temperature was 20,000° (hotter than any known star), in which case we should have detected its error. If we cannot feel confident that the answers of the theory are true, it must be admitted that it has shown some discretion in lying without being found out.

It would not be surprising if individual stars occasionally depart considerably from the calculated results, because at present no serious attempt has been made to take into account rotation, which may modify the conditions when sufficiently rapid. That appears to be the next step needed for a more exact study of the question.

Probably the greatest need of stellar astronomy at the present day, in order to make sure that our theoretical deductions are starting on the right lines, is some means of measuring the apparent angular diameters of stars. At present we can calculate them approximately from theory, but there is no observational check. We believe we know with fair accuracy the apparent surface brightness corresponding to each spectral type; then all that is necessary is to divide the total apparent brightness by this surface brightness, and the result is the angular area subtended by the star. The unknown distance is not involved, because surface brightness is independent of distance. Thus the estimation of the angular diameter of any star seems to be a very simple matter. For instance, the star with the greatest apparent diameter is almost certainly Betelgeuse, diameter 0·051″. Next to it comes Antares, 0·043″. Other examples are Aldebaran 0·022″, Arcturus 0·020″, Pollux 0·013″. Sirius comes rather low down with diameter 0·007″. The following table may be of interest as showing the angular diameters expected for stars of various types and visual magnitudes :—

Probable Angular Diameters of Stars.

Vis. Mag.	A	F	G	K	M
m	"	"	"	"	"
0·0	0·0034	0·0054	0·0098	0·0219	0·0859
2·0	0·0014	0·0022	0·0039	0·0087	0·0342
4·0	0·0005	0·0009	0·0016	0·0035	0·0136

However confidently we may believe in these values, it would be an immense advantage to have this first step in our deductions placed beyond doubt. If the direct measurement of these diameters could be ,made with any accuracy it would make a wonderfully rapid advance in our knowledge. The prospects of accomplishing some part of this task are now quite hopeful. We have learnt with great interest this year that work is being carried out by interferometer methods with the 100-in. reflector at Mount Wilson, and the results are most promising. At present the method has been applied only to measuring the separation of close double stars, but there seems to be no doubt that an angular diameter of 0·05″ is well within reach. Although the great mirror is used for convenience, the interferometer method does not in principle require great apertures, but rather two small apertures widely separated, as in a range-finder. Prof. Hale has stated, moreover, that successful results were obtained on nights of poor seeing. Perhaps it would be unsafe to assume that "poor seeing" at Mount Wilson means quite the same thing as it does for us, and I anticipate that atmospheric disturbance will ultimately set the limit to what can be accomplished. But even if we have to send special expeditions to the top of one of the highest mountains in the world, the attack on this far-reaching problem must not be allowed to languish.

I spoke earlier of the radiation-pressure exerted by the outflowing heat, which has an important effect on the equilibrium of a star. It is quite easy to calculate what proportion of the weight of the material is supported in this way; it depends on neither the density nor the opacity, but solely on the star's total mass and on the molecular weight. No astronomical data are needed; the calculation involves only fundamental physical constants found by laboratory researches. Here are the figures, first for average molecular weight 3·0 :—

> For mass $\frac{1}{2}$×sun, fraction of weight supported by radiation-pressure=0·044.
> For mass 5×sun, fraction of weight supported by radiation-pressure=0·457.

For molecular weight 5·0 the corresponding fractions are 0·182 and 0·645.

The molecular weight can scarcely go beyond this range,[2] and for the conclusions I am about to draw it does not much matter which limit we take. Probably 90 per cent. of the giant stars have masses between $\frac{1}{2}$ and 5 times the sun's, and we see that this is just the range in which radiation-pressure rises from unimportance to importance. It seems clear that a globe of gas of larger mass, in which radiation-pressure and gravitation are nearly balancing, would be likely to be unstable. The condition may not be strictly unstable in itself, but a small rotation or perturbation would make it so. It may therefore be conjectured that, if nebulous material began to concentrate into a mass much greater than five times the sun's, it would probably break up, and continue to redivide until more stable masses resulted. Above the upper limit the chances of survival are small; when the lower limit is approached the danger has practically disappeared, and there is little likelihood of any further breaking-up. Thus the final masses are left distributed almost entirely between the limits given. To put the matter slightly differently, we are able to predict from general principles that the material of the stellar universe will aggregate primarily into masses chiefly lying between 10^{33} and 10^{34} grams; and this is just the

2 As an illustration of these limits, iron has 26 outer electrons; if 13 break away the average molecular weight is 5; if 18 break away the molecular weight is 3. Eggert (*Phys. Zeits.*, 1919, p. 570) has suggested by thermodynamical reasoning that in most cases the two outer rings (16 electrons) would break away in the stars. The comparison of theory and observation for the dwarf stars also points to a molecular weight a little greater than 3.

magnitude of the masses of the stars according to astronomical observation.[3]

This study of the radiation and internal conditions of a star brings forward very pressingly a problem often debated in this Section : What is the source of the heat which the sun and stars are continually squandering? The answer given is almost unanimous —that it is obtained from the gravitational energy converted as the star steadily contracts. But almost as unanimously this answer is ignored in its practical consequences. Lord Kelvin showed that this hypothesis, due to Helmholtz, necessarily dates the birth of the sun about 20,000,000 years ago; and he made strenuous efforts to induce geologists and biologists to accommodate their demands to this time-scale. I do not think they proved altogether tractable. But it is among his own colleagues, physicists and astronomers, that the most outrageous violations of this limit have prevailed. I need only refer to Sir George Darwin's theory of the earth-moon system, to the present Lord

Rayleigh's determination of the age of terrestrial rocks from occluded helium, and to all modern discussions of the statistical equilibrium of the stellar system. No one seems to have any hesitation, if it suits him, in carrying back the history of the earth long before the supposed date of formation of the solar system; and, in some cases at least, this appears to be justified by experimental evidence which it is difficult to dispute. Lord Kelvin's date of the creation of the sun is treated with no more respect than Archbishop Ussher's.

The serious consequences of this contraction hypothesis are particularly prominent in the case of giant stars, for the giants are prodigal with their heat and radiate at least a hundred times as fast as the sun. The supply of energy which suffices to maintain the sun for 10,000,000 years would be squandered by a giant star in less than 100,000 years. The whole evolution in the giant stage would have to be very rapid. In 18,000 years at the most a typical star must pass from the initial M stage to type G. In 80,000 years it has reached type A near the top of the scale, and is about to start on the downward path. Even these figures are probably very much over-estimated.[4] Most of the naked-eye stars are still in the giant stage. Dare we believe that they were all formed within the last 80,000 years? The telescope reveals to us objects remote not only in distance, but also in time. We can turn it on a globular cluster and behold what was passing 20,000, 50,000, even 200,000 years ago unfortunately not all in the same cluster, but in different clusters representing different epochs of the past. As Shapley has pointed out, the verdict appears to be "no change." This is perhaps not conclusive, because it does not follow that individual stars have suffered no change in the interval; but it is difficult to resist the impression that the evolution of the stellar universe proceeds at a slow, majestic pace, with respect to which these periods of time are insignificant.

There is another line of astronomical evidence which appears to show more definitely that the evolution of the stars proceeds far more slowly than the contraction hypothesis allows; and perhaps it may ultimately enable us to measure the true rate of progress. There are certain stars, known as Cepheid variables, which undergo a regular fluctuation of light of a characteristic

[3] By admitting plausible assumptions closer limits could be drawn. Taking the molecular weight as 3'5, and assuming that the most critical condition is when. $\frac{1}{8}$ of gravitation is counterbalanced (by analogy with the case of rotating spheroids, in which centrifugal force opposes gravitation and creates instability), we find that the critical mass is just twice that of the sun, and stellar masses may be expected to cluster closely round this value.

[4] I have taken the ratio of specific heats at the extreme possible value, $\frac{5}{3}$; that is to say, no allowance has been made for the energy needed for ionisation and internal vibrations of the atoms, which makes a further call on the scanty supply available.

세상에서 가장 쉬운 과학 수업 별의 물리학

kind, generally with a period of a few days. This light change is *not* due to eclipse. Moreover, the colour quality of the light changes between maximum and minimum, evidently pointing to a periodic change in the physical condition of the star. Although these objects were formerly thought to be double stars, it now seems clear that this was a misinterpretation of the spectroscopic evidence. There is, in fact, no room for the hypothetical companion star; the orbit is so small that we should have to place it inside the principal star. Everything points to the period of the light pulsation being something intrinsic in the star; and the hypothesis advocated by Shapley, that it represents a mechanical pulsation of the star, seems to be the most plausible. I have already mentioned that the observed period does, in fact, agree with the calculated period of mechanical pulsation, so that the pulsation explanation survives one fairly stringent test. But whatever the cause of the variability, whether pulsation or rotation, provided only that it is intrinsic in the star, and not forced from outside, the density must be the leading factor in determining the period. If the star is contracting so that its density changes appreciably, the period cannot remain constant. Now, on the contraction hypothesis the change of density must amount to at least 1 per cent. in forty years. (I give the figures for δ Cephei, the best-known variable of this class.) The corresponding change of period should be very easily detectable. For δ Cephei the period ought to decrease 40 seconds annually.

Now δ Cephei has been under careful observation since 1785, and it is known that the change of period, if any, must be very small. S. Chandler found a decrease of period of $\frac{1}{20}$ second per annum, and in a recent investigation E. Hertzsprung has found a decrease of $\frac{1}{10}$ second per annum. The evidence that there is any decrease at all rests almost entirely on the earliest observations made before 1800, so that it is not very certain; but in any case the evolution is proceeding at not more than $\frac{1}{400}$ of the rate required by the contraction hypothesis. There must at this stage of the evolution of the star be some other source of energy which prolongs the life of the star 400-fold. The time-scale so enlarged would suffice for practically all reasonable demands.

I hope the dilemma is plain. Either we must admit that whilst the density changes 1 per cent. a certain period intrinsic in the star can change no more than $\frac{1}{800}$ of 1 per cent., or we must give up the contraction hypothesis.

If the contraction theory were proposed to-day as a novel hypothesis I do not think it would stand the smallest chance of acceptance. From all sides—biology, geology, physics, astronomy—it would be objected that the suggested source of energy was hopelessly inadequate to provide the heat spent during the necessary time of evolution; and, so far as it is possible to interpret observational evidence confidently, the theory would be held to be negatived definitely. Only the inertia of tradition keeps the contraction hypothesis alive—or, rather, not alive, but an unburied corpse. But if we decide to inter the corpse, let us frankly recognise the position in which we are left. A star is drawing on some vast reservoir of energy by means unknown to us. This reservoir can scarcely be other than the sub-atomic energy which, it is known, exists abundantly in all matter; we sometimes dream that man will one day learn how to release it and use it for his service. The store is well-nigh inexhaustible, if only it could be tapped. There is sufficient in the sun to maintain its output of heat for 15 billion years.

Certain physical investigations in the past year, which I hope we may hear about at this meeting, make it probable to my mind that some portion of this sub-atomic energy is actually being set free in the stars. F. W. Aston's experiments seem to leave no room for doubt that all the elements are constituted out of hydrogen atoms bound together with negative electrons. The nucleus of the helium atom, for example, consists of four hydrogen atoms bound with two electrons. But Aston has further shown conclusively that the mass of the helium atom is less than the sum of the masses of the four hydrogen atoms which enter into it; and in this, at any rate, the chemists agree with him. There is a loss of mass in the synthesis amounting to about 1 part in 120, the atomic weight of hydrogen being 1·008 and that of helium just 4. I will not dwell on his beautiful proof of this, as you will, no doubt, be able to hear it from himself. Now mass cannot be annihilated, and the deficit can only represent the mass of the electrical energy set free in the transmutation. We can therefore at once calculate the quantity of energy liberated when helium is made out of hydrogen. If 5 per cent. of a star's mass consists initially of hydrogen atoms, which are gradually being combined to form more complex elements, the total heat liberated will more than suffice for our demands, and we need look no further for the source of a star's energy.

But is it possible to admit that such a transmutation is occurring? It is difficult to assert, but perhaps more difficult to deny, that this is going on. Sir Ernest Rutherford has recently been breaking down the atoms of oxygen and nitrogen, driving out an isotope of helium from them; and what is possible in the Cavendish Laboratory may not be too difficult in the sun. I think that the suspicion has been generally entertained that the stars are the crucibles in which the lighter atoms which abound in the nebulæ are compounded into more complex elements. In the stars matter has its preliminary brewing to prepare the greater variety of elements which are needed for a world of life. The radio-active elements must have been formed at no very distant date; and their synthesis, unlike the generation of helium from hydrogen, is endothermic. If combinations requiring the addition of energy can occur in the stars, combinations which liberate energy ought not to be impossible.

We need not bind ourselves to the formation of helium from hydrogen as the sole reaction which supplies the energy, although it would seem that the further stages in building up the elements involve much less liberation, and sometimes even absorption, of energy. It is a question of accurate measurement of the deviations of atomic weights from integers, and up to the present hydrogen is the only element for which Dr. Aston has been able to detect the deviation. No doubt we shall learn more about the possibilities in due time. The position may be summarised in these terms: the atoms of all elements are built of hydrogen atoms bound together, and presumably have at one time been formed from hydrogen; the interior of a star seems as likely a place as any for the evolution to have occurred; whenever it did occur a great amount of energy must have been set free; in a star a vast quantity of energy is being set free which is hitherto unaccounted for. You may draw a conclusion if you like.

If, indeed, the sub-atomic energy in the stars is being freely used to maintain their great furnaces, it seems to bring a little nearer to fulfilment our dream of controlling this latent power for the well-being of the human race—or for its suicide.

So far as the immediate needs of astronomy are concerned, it is not of any great consequence whether in this suggestion we have actually laid a finger on the true source of the heat. It is sufficient if the discussion opens our eyes to the wider possibilities. We can get rid of the obsession that there is no other conceivable supply besides contraction, but we need not again cramp ourselves by adopting prematurely what is perhaps a still wilder guess. Rather we should admit that the source is not certainly known, and seek for any possible astronomical evidence which may help to define its necessary character. One piece of evidence of this kind may be worth mentioning. It seems clear that it must be the high temperature inside the stars which determines the liberation of energy, as H. N. Russell has pointed out (Pubns. Ast. Soc. Pacific, August, 1919). If so, the supply may come mainly from the hottest region at the centre. I have already stated that the general uniformity of the opacity of the stars is much more easily intelligible if it depends on scattering rather than on true absorption; but it did not seem possible to reconcile the deduced stellar opacity with the theoretical scattering coefficient. Within reasonable limits it makes no great difference in our calculations at what parts of the star the heat energy is supplied, and it was assumed that it comes more or less evenly from all parts, as would be the case on the contraction theory. The possibility was scarcely contemplated that the energy is supplied entirely in a restricted region round the centre. Now, the more concentrated the supply, the lower is the opacity requisite to account for the observed radiation. I have not made any detailed calculations, but it seems possible that for a sufficiently concentrated source the deduced and the theoretical coefficients could be made to agree, and there does not seem to be any other way of accomplishing this. Conversely, we might perhaps argue that the present discrepancy of the coefficients shows that the energy supply is not spread out in the way required by the contraction hypothesis, but belongs to some new source available only at the hottest, central part of the star.

I should not be surprised if it is whispered that this address has at times verged on being a little bit speculative; perhaps some outspoken friend may bluntly say that it has been highly speculative from beginning to end. I wonder what is the touchstone by which we may test the legitimate development of scientific theory and reject the idly speculative. We all know of theories which the scientific mind instinctively rejects as fruitless guesses; but it is difficult to specify their exact defect or to supply a rule which will show us when we ourselves do err. It is often supposed that to speculate and to make hypotheses are the same thing; but more often they are opposed. It is when we let our thoughts stray outside venerable, but sometimes insecure, hypotheses that we are said to speculate. Hypothesis limits speculation. Moreover, distrust of speculation often serves as a cover for loose thinking; wild ideas take anchorage in our minds and influence our outlook; whilst it is considered too speculative to subject them to the scientific scrutiny which would exorcise them.

If we are not content with the dull accumulation of experimental facts, if we make any deductions or generalisations, if we seek for any theory to guide us, some degree of speculation cannot be avoided. Some will prefer to take the interpretation which seems to be indicated most immediately and at once adopt that as an hypothesis; others will rather seek to explore and classify the widest possibilities which are not definitely inconsistent with the facts. Either choice has it dangers : the first may be too narrow a view and lead progress into a cul-de-sac; the second may be so broad that it is useless as a guide, and diverges indefinitely from experimental knowledge. When this last case happens, it must be concluded that the knowledge is not yet ripe for theoretical treatment and that speculation is premature. The time when speculative theory and observational research may profitably go hand in hand is when the possibilities, or at any rate the probabilities, can be narrowed down by experiment, and the theory can indicate the tests by which the remaining wrong paths may be blocked up one by one.

The mathematical physicist is in a position of peculiar difficulty. He may work out the behaviour of an ideal model of material with specifically defined properties, obeying mathematically exact laws, and so far his work is unimpeachable. It is no more speculative than the binomial theorem. But when he claims a serious interest for his toy, when he suggests that his model is like something going on in Nature, he inevitably begins to speculate. Is the actual body really like the ideal model? May not other unknown conditions intervene? He cannot be sure, but he cannot suppress the comparison; for it is by looking continually to Nature that he is guided in his choice of a subject. A common fault, to which he must often plead guilty, is to use for the comparison data over which the more experienced observer shakes his head; they are too insecure to build extensively upon. Yet even in this, theory may help observation by showing the kind of data which it is especially important to improve.

I think that the more idle kinds of speculation will be avoided if the investigation is conducted from the right point of view. When the properties of an ideal model have been worked out by rigorous mathematics, all the underlying assumptions being clearly understood, then it becomes possible to say that such-and-such properties and laws lead precisely to such-and-such effects. If any other disregarded factors are present, they should now betray themselves when a comparison is made with Nature. There is no need for disappointment at the failure of the model to give perfect agreement with observation; it has served its purpose, for it has distinguished what are the features of the actual phenomena which require new conditions for their explanation. A general preliminary agreement with observation is necessary, otherwise the model is hopeless; not that it is necessarily wrong so far as it goes, but it has evidently put the less essential properties foremost. We have been pulling at the wrong end of the tangle, which has to be unravelled by a different approach. But after a general agreement with observation is established, and the tangle begins to loosen, we should always make ready for the next knot. I suppose that the applied mathematician whose theory has just passed one still more stringent test by observation ought not to feel satisfaction, but rather disappointment—"Foiled again! This time I *had* hoped to find a discordance which would throw light on the points where my model could be improved." Perhaps that is a counsel of perfection; I own that I have never felt very keenly a disappointment of this kind.

Our model of Nature should not be like a building —a handsome structure for the populace to admire, until in the course of time someone takes away a corner-stone and the edifice comes toppling down. It should be like an engine with movable parts. We need not fix the position of any one lever; that is to be adjusted from time to time as the latest observations indicate. The aim of the theorist is to know the train of wheels which the lever sets in motion—that binding of the parts which is the soul of the engine.

In ancient days two aviators procured to themselves wings. Dædalus flew safely through the middle air across the sea, and was duly honoured on his landing. Young Icarus soared upwards towards the sun until the wax which bound his wings melted, and his flight ended in fiasco. In weighing their achievements perhaps there is something to be said for Icarus. The classic authorities tell us that he was only "doing a stunt," but I prefer to think of him as the man who certainly brought to light a constructional defect in the flying-machines of his day. So, too, in science. Cautious Dædalus will apply his theories where he feels most confident they will safely go; but by his excess of caution their hidden weaknesses cannot be brought to light. Icarus will strain his theories to the breaking-point until the weak joints gape. For a

세상에서 가장 쉬운 과학 수업 별의 물리학

spectacular stunt? Perhaps partly; he is often very human. But if he is not yet destined to reach the sun and solve for all time the riddle of its constitution, yet he may hope to learn from his journey some hints to build a better machine.

논문 웹페이지

MARCH 1, 1939   PHYSICAL REVIEW   VOLUME 55

Energy Production in Stars*

H. A. Bethe

Cornell University, Ithaca, New York

(Received September 7, 1938)

It is shown that the *most important source of energy in ordinary stars is the reactions of carbon and nitrogen with protons*. These reactions form a cycle in which the original nucleus is reproduced, *viz*. $C^{12}+H=N^{13}$, $N^{13}=C^{13}+\epsilon^+$, $C^{13}+H=N^{14}$, $N^{14}+H=O^{15}$, $O^{15}=N^{15}+\epsilon^+$, $N^{15}+H=C^{12}+He^4$. Thus carbon and nitrogen merely serve as catalysts for the combination of four protons (and two electrons) into an α-particle (§7).

The carbon-nitrogen reactions are unique in their cyclical character (§8). For all nuclei lighter than carbon, reaction with protons will lead to the emission of an α-particle so that the original nucleus is permanently destroyed. For all nuclei heavier than fluorine, only radiative capture of the protons occurs, also destroying the original nucleus. Oxygen and fluorine reactions mostly lead back to nitrogen. Besides, these heavier nuclei react much more slowly than C and N and are therefore unimportant for the energy production.

The agreement of the carbon-nitrogen reactions with observational data (§7, 9) is excellent. In order to give the correct energy evolution in the sun, the central temperature of the sun would have to be 18.5 million degrees while integration of the Eddington equations gives 19. For the brilliant star Y Cygni the corresponding figures are 30 and 32. This good agreement holds for all bright stars of the main sequence, but, of course, not for giants.

For fainter stars, with lower central temperatures, the reaction $H+H=D+\epsilon^+$ and the reactions following it, are believed to be mainly responsible for the energy production. (§10)

It is shown further (§5–6) that *no elements heavier than He^4 can be built up in ordinary stars*. This is due to the fact, mentioned above, that all elements up to boron are disintegrated by proton bombardment (α-emission!) rather than built up (by radiative capture). The instability of Be^8 reduces the formation of heavier elements still further. The production of neutrons in stars is likewise negligible. The heavier elements found in stars must therefore have existed already when the star was formed.

Finally, the suggested mechanism of energy production is used to draw conclusions about astrophysical problems, such as the mass-luminosity relation (§10), the stability against temperature changes (§11), and stellar evolution (§12).

---

§1. INTRODUCTION

THE progress of nuclear physics in the last few years makes it possible to decide rather definitely which processes can and which cannot occur in the interior of stars. Such decisions will be attempted in the present paper, the discussion being restricted primarily to main sequence stars. The results will be at variance with some current hypotheses.

The first main result is that, under present conditions, no elements heavier than helium can be built up to any appreciable extent. Therefore we must assume that the heavier elements were built up *before* the stars reached their present state of temperature and density. No attempt will be made at speculations about this previous state of stellar matter.

The energy production of stars is then due entirely to the combination of four protons and two electrons into an α-particle. This simplifies the discussion of stellar evolution inasmuch as the amount of heavy matter, and therefore the opacity, does not change with time.

The combination of four protons and two electrons can occur essentially only in two ways. The first mechanism starts with the combination of two protons to form a deuteron with positron emission, *viz*.

$$H+H=D+\epsilon^+. \tag{1}$$

The deuteron is then transformed into He^4 by further capture of protons; these captures occur very rapidly compared with process (1). The second mechanism uses carbon and nitrogen as catalysts, according to the chain reaction

$$\begin{aligned}
C^{12}+H &= N^{13}+\gamma, & N^{13} &= C^{13}+\epsilon^+ \\
C^{13}+H &= N^{14}+\gamma, & & \\
N^{14}+H &= O^{15}+\gamma, & O^{15} &= N^{15}+\epsilon^+ \\
N^{15}+H &= C^{12}+He^4. &
\end{aligned} \tag{2}$$

The catalyst C^{12} is reproduced in all cases except about one in 10,000, therefore the abundance of carbon and nitrogen remains practically unchanged (in comparison with the change of the number of protons). The two reactions (1) and

* Awarded an A. Cressy Morrison Prize in 1938, by the New York Academy of Sciences.

세상에서 가장 쉬운 과학 수업 별의 물리학

(2) are about equally probable at a temperature of $16 \cdot 10^6$ degrees which is close to the central temperature of the sun ($19 \cdot 10^6$ degrees[1]). At lower temperatures (1) will predominate, at higher temperatures, (2).

No reaction other than (1) or (2) will give an appreciable contribution to the energy production at temperatures around $20 \cdot 10^6$ degrees such as are found in the interior of ordinary stars. The lighter elements (Li, Be, B) would "burn" in a very short time and are not replaced as is carbon in the cycle (2), whereas the heavier elements (O, F, etc.) react too slowly. Helium, which is abundant, does not react with protons because the product, Li^5, does not exist; in fact, the energy evolution in stars can be used as a strong additional argument against the existence of He^5 and Li^5 (§3).

Reaction (2) is sufficient to explain the energy production in very luminous stars of the main sequence as Y Cygni (although there are difficulties because of the quick exhaustion of the energy supply in such stars which would occur on any theory, §9). Neither of the reactions (1) or (2) is capable of accounting for the energy production in giants; if nuclear reactions are at all responsible for the energy production in these stars it seems that the only ones which could give sufficient energy are

$$H^2 + H = He^3 \quad (3)$$

$$Li^{6, 7} + H = He^{3, 4} + He^4.$$

It seems, however, doubtful whether the energy production in giants is due to nuclear reactions at all.[2]

We shall first calculate the energy production by nuclear reactions (§2, 4). Then we shall prove the impossibility of building up heavier elements under existing conditions (§5–6). Next we shall discuss the reactions available for energy production (§5, 7) and the results will be compared with available material on stellar temperatures and densities (§8, 9). Finally, we shall discuss the astrophysical problems of the mass-luminosity relation (§10), the stability of stars against temperature changes (§11) and stellar evolution (§12).

§2. FORMULA FOR ENERGY PRODUCTION

The probability of a nuclear reaction in a gas with a Maxwellian velocity distribution was first calculated by Atkinson and Houtermans.[3] Recently, an improved formula was derived by Gamow and Teller.[4] The total number of processes per gram per second is[4]

$$p = \frac{4}{3^{5/2}} \frac{\rho x_1 x_2}{m_1 m_2} \frac{\Gamma}{\hbar} - a R^2 e^{4(2R/a)^{\frac{1}{2}}} \tau^2 e^{-\tau}. \quad (4)$$

Here ρ is the density of the gas, $x_1 x_2$ the concentrations (by weight) of the two reacting types of nuclei, $m_1 m_2$ their masses, $Z_1 e$ and $Z_2 e$ their charges, $m = m_1 m_2/(m_1 + m_2)$ the reduced mass, R the combined radius,

$$a = \hbar^2/me^2 Z_1 Z_2 \quad (5)$$

the "Bohr radius" for the system, $\Gamma/\hbar$ the probability of the nuclear reaction, in sec.$^{-1}$, after penetration, and

$$\tau = 3 \left(\frac{\pi^2 m e^4 Z_1^2 Z_2^2}{2 \hbar^2 k T} \right)^{\frac{1}{3}}. \quad (6)$$

If we measure ρ in g/cm^3, Γ in volts and T in units of 10^6 degrees, we have

$$p = 5.3 \cdot 10^{25} \rho x_1 x_2 \Gamma \varphi(Z_1, Z_2) \tau^2 e^{-\tau} \quad g^{-1} sec.^{-1}, \quad (7)$$

$$\tau = 42.7 (Z_1 Z_2)^{\frac{2}{3}} (A/T)^{\frac{1}{3}}, \quad (8)$$

$$\varphi = \frac{1}{A_1 A_2 (Z_1 Z_2 A)^{\frac{1}{3}}} \left(\frac{8R}{a} \right)^2 e^{2(8R/a)^{\frac{1}{2}}}. \quad (9)$$

where $A_1 A_2$ are the atomic weights of the reacting nuclei ($A_i = m_i/m_H$), $A = m/m_H$, m_H = mass of hydrogen. For the combined radius of nuclei 1 and 2 we put

$$R = 1.6 \cdot 10^{-13} (A_1 + A_2)^{\frac{1}{3}} cm. \quad (10)$$

Then we obtain for φ the values given in Table I. The values of φ for isotopes of the same element differ only very slightly.

The values of Γ for reactions giving particles can be deduced from the *observed* cross sections

[1] B. Strömgren, Ergebn. d. Exakt. Naturwiss. 16, 465 (1937).

[2] G. Gamow, private communication.

[3] R. d'E. Atkinson and F. G. Houtermans, Zeits. f. Physik 54, 656 (1929).

[4] G. Gamow and E. Teller, Phys. Rev. 53, 608 (1938), Eq. (3).

TABLE I. *Values of φ for various nuclear reactions.*

REACTION	$\dfrac{R}{(10^{-13}\text{ CM})}$	φ	REACTION	R	φ
H^2+H	2.3	0.38	$Si^{30}+H$	5.0	29.3
H^3+H	2.5_5	0.48	$Cl^{37}+H$	5.4	75
He^4+H	2.7_5	0.81	H^2+H^2	2.5_5	0.67
Li^7+H	3.2	0.91	Be^7+H^2	3.3	1.18
Be^9+H	3.4_5	1.16	Be^7+He^3	3.4_5	7.9
$B^{11}+H$	3.6_5	1.52	He^4+H^2	2.9	0.57
$C^{12}+H$	3.7_5	2.00	He^4+He^3	3.0_5	1.09
$N^{14}+H$	3.9_5	2.78	He^4+He^4	3.2	1.29
$O^{16}+H$	4.1	3.80	Li^7+He^4	3.5_5	4.9
$F^{19}+H$	4.3_5	5.5	Be^7+He^4	3.5_5	13.2
$Ne^{22}+H$	4.5_5	7.7	Be^8+He^4	3.6_5	16.2
$Mg^{26}+H$	4.8	13.2	$C^{12}+He^4$	4.0	230

TABLE II. *Cross sections and widths for some nuclear reactions giving particles.*

REACTION	REF.	$\dfrac{E}{\text{KV}}$	$\dfrac{\sigma}{\text{CM}^2}$	$\dfrac{R}{\text{CM}}$	Γ
$H^2+H^2=He^3+n^1$	5	100	$1.7\cdot10^{-26}$	$2.6\cdot10^{-13}$	$3\cdot10^5$
$Li^7+H^1=2He^4$	6	42	$1.7\cdot10^{-30}$	$3.2\cdot10^{-13}$	$4\cdot10^4$
$Li^6+H^1=He^4+He^3$	yield	$\sim$ same as Li^7+H in natural Li target			$5\cdot10^5$
$Li^6+H^2=\begin{cases}2He^4\\Li^7+H\end{cases}$	7	212	$1.9\cdot10^{-26}$	$3.2\cdot10^{-13}$	$4\cdot10^6$
$Li^7+H^2=2He^4+n$	7	212	$5.5\cdot10^{-26}$	$3.3\cdot10^{-13}$	10^7
$Be^9+H^1=\begin{cases}Li^6+He^4\\Be^8+H^2\end{cases}$	7a	212	$1.1\cdot10^{-25}$	$3.5\cdot10^{-13}$	$1.7\cdot10^7$
$Be^9+H^2=\begin{cases}Li^7+He^4\\Be^8+H^3\\Be^{10}+H^1\end{cases}$	7a	212	$5\cdot10^{-26}$	$3.6\cdot10^{-13}$	$6\cdot10^6$
$B^{11}+H^1=3He^4$	7b	212	$6\cdot10^{-28}$	$3.7\cdot10^{-13}$	$2\cdot10^6$

of such reactions with the use of the formula (cf. reference 4, Eq. (2))

$$\sigma = \frac{\pi R^2}{2E}\frac{A_1+A_2}{A_2}\Gamma\exp\left[\left(\frac{32R}{a}\right)^{\frac{1}{2}} - \frac{2\pi e^2}{\hbar v}Z_1Z_2\right] \quad (11)$$

where E is the absolute energy of the incident particle (particle 1). Table II gives the experimental results for some of the better investigated reactions. In each case, experiments with low energy particles were chosen in order to make the conditions as similar as possible to those in stars where the greatest number of nuclear reactions is due to particles of about 20 kilovolts energy. The cross sections were in each case calculated from the thick target yield with the help of the range-energy relation of Herb,

[5] R. Ladenburg and M. H. Kanner, Phys. Rev. **52**, 911 (1937).
[6] H. D. Doolittle, Phys. Rev. **49**, 779 (1936).
[7] J. H. Williams, W. G. Shepherd and R. O. Haxby, Phys. Rev. **52**, 390 (1937).
[7a] J. H. Williams, R. O. Haxby and W. G. Shepherd, Phys. Rev. **52**, 1031 (1937).
[7b] J. H. Williams, W. H. Wells, J. T. Tate and E. J. Hill, Phys. Rev. **51**, 434 (1937).

Bellamy, Parkinson and Hudson.[8] The widths obtained (last column of Table II) are mostly between $3\cdot10^5$ and $2\cdot10^7$ ev, with the exception of the reaction $Li^7+H=2He^4$ which is known to be "improbable."[9]

The γ-ray widths Γ_γ can be obtained from observed resonance capture of protons. Table III gives the experimental results. Two of the older data were taken from Table XXXIX of reference 10; all the others are from more recent experiments on proton[11-14] and neutron[15] capture. Although the results of different investigators differ considerably (e.g., for $Li^7+H^1=Be^8$, Γ is between 4 and 40 volts the latter value being more likely) they seem to lie generally between about $\frac{1}{2}$ and 40 volts. Ordinarily, the width is somewhat larger for the more energetic γ-rays, as is expected theoretically. A not too bad approximation to the experiments is obtained by using the theoretical formula for dipole radiation (reference 10, Eq. (711b)) with an oscillator strength of 1/50. This gives

$$\Gamma_\gamma \sim 0.1 E_\gamma^2, \quad (12)$$

TABLE III. *γ-ray widths of nuclear levels.*

REACTION	REFERENCE	WIDTH (VOLTS)	γ-RAY ENERGY (MEV)
$Li^7+H^1=Be^8+\gamma$	$\begin{cases}10\\11\end{cases}$	$\begin{array}{c}4\\40\end{array}$	17
$B^{11}+H^1=C^{12}+\gamma$	11	0.6	12, 16
$C^{12}+H^1=N^{13}+\gamma$	12	0.6	2
$C^{13}+H^1=N^{14}+\gamma$	13, 14	30	4, 8
$F^{19}+H^1=Ne^{20}+\gamma$	10	0.6, 8, 18	6
$C^{12}+n^1=C^{13}+\gamma$	15	<2.5	5
$O^{16}+n^1=O^{17}+\gamma$	15	<2.5	4

[8] D. B. Parkinson, R. G. Herb, J. C. Bellamy and C. M. Hudson, Phys. Rev. **52**, 75 (1937).
[9] M. Goldhaber, Proc. Camb. Phil. Soc. **30**, 560 (1934).
[10] H. A. Bethe, Rev. Mod. Phys. **9**, 71 (1937).
[11] W. A. Fowler, E. R. Gaerttner and C. C. Lauritsen, Phys. Rev. **53**, 628 (1938).
[12] R. B. Roberts and N. P. Heydenburg, Phys. Rev. **53**, 374 (1938).
[13] P. I. Dee, S. C. Curran and V. Petržilka, Nature **141**, 642 (1938). The γ-rays from C^{13} give about equally as many counts as those from C^{12}. The efficiency of the counter for C^{13} γ-rays is about twice that for C^{12}, because the cross section for production of Compton and pair electrons is smaller by a factor 2/3 while the range of these electrons is about 3 times longer. With an abundance of C^{13} of about 1 percent, the γ-width for this nucleus becomes 50 times that of C^{12}. I am indebted to Dr. Rose for these calculations.
[14] M. E. Rose, Phys. Rev. **53**, 844 (1938).
[15] O. R. Frisch, H. v. Halban and J. Koch, Nature **140**, 895 (1937); Danish Acad. Sci. **15**, 10 (1937).

where E_γ is the γ-ray energy in mMU (milli-mass-units), and Γ_γ the γ-ray width in volts. For quadrupole radiation, theory gives about

$$\Gamma_\gamma \sim 5 \cdot 10^{-4} E_\gamma{}^4 \quad \text{(quadrupole)}. \quad (12a)$$

Formulae (12), and (12a) will be used in the calculations where experimental data are not available; they may, in any individual case, be in error by a factor 10 or more but such a factor is not of great importance compared with other uncertainties.

It should be noted that quite generally radiative processes are rare compared with particle emission. According to the figures given in Tables II and III, the ratio of probabilities is 10^4–10^5 in favor of particle reactions.

In a number of cases, the reaction of a nucleus A with a heavy particle (proton, alpha-) must compete with natural β-radioactivity of A or with electron capture. In those cases where the lifetime of radioactive nuclei is not known experimentally, we use the Fermi theory. According to this theory, the decay constant for β-emission is[16]

$$\beta = 0.9 \cdot 10^{-4} f(W) |G|^2 \ \text{sec.}^{-1}. \quad (13)$$

The matrix element G is about unity for strongly allowed transitions, and

$$f(W) = (W^2 - 1)^{\frac{1}{2}} \left(\frac{1}{30} W^4 - \frac{3}{20} W^2 - \frac{2}{15} \right)$$
$$+ \tfrac{1}{4} W \log \{ W + (W^2 - 1)^{\frac{1}{2}} \}, \quad (13a)$$

where W is the maximum energy of the β-particle, including its rest mass, in units of mc^2 (m = electron mass).

The probability of electron capture is

$$\beta_C = 0.9 \cdot 10^{-4} \pi^2 N (\hbar/mc)^3 W^2 |G|^2 \ \text{sec.}^{-1}, \quad (14)$$

where W is the energy of the emitted neutrino in units of mc^2 and N the number of electrons per unit volume. If the hydrogen concentration is x_H, we have (reference 1, p. 482) $N = 6 \cdot 10^{23} \rho \cdot \frac{1}{2}(1 + x_H)$ (ρ the density), and

$$\beta_C = 1.5 \cdot 10^{-11} \rho (1 + x_H) W^2 |G|^2 \ \text{sec.}^{-1}. \quad (14a)$$

---

[16] C. L. Critchfield and H. A. Bethe, Phys. Rev. 54, 248, 862 (L) (1938).

§3. Stability of Unknown Isotopes

For the discussion of nuclear reactions it is essential to know whether or not certain isotopes exist (such as Li^4, Li^5, Be^6, Be^8, B^8, B^9, C^{10}, etc.). The criterion for the existence of a nucleus is its energetic stability against spontaneous disintegration into heavy particles (emission of a neutron, proton or alpha-particle). Whenever a light nucleus is energetically unstable against heavy particle emission, its life will be a very small fraction of a second (usually $\sim 10^{-20}$ sec.) even if the instability is slight (e.g., Be^8 will have a life of 10^{-13} sec. if it is by 50 kv heavier than two α-particles[17]).

For the question of the lifetime of radioactive nuclei, it is also necessary to know the mass difference between isobars. Similar information is required for estimating the γ-ray width in capture reactions (cf. Eqs. (12), (12a)).

H^3 and He^3

The most recent determination[18] of the reaction energy in the reaction $H^2 + H^2 = He^3 + n^1$ yielded 3.29 Mev as compared with 3.98 in $H^2 + H^2 = H^3 + H^1$. With a mass difference of 0.75 Mev between neutron and hydrogen atom,[19] this makes He^3 more stable than H^3 by 0.06 Mev. This would be in agreement with the experimental fact that no H^3 is present in natural hydrogen to more than 1 part in 10^{12}. Even if He^3 should turn out to be heavier than H^3, the difference cannot be greater than about 0.05 Mev $= 0.1 \ mc^2$ which would make the life of He^3 exceedingly long (~ 2000 years at the center of the sun, (Eq. (14a)), 3000 years in the complete atom, on earth).

H^4 and Li^4

As was first pointed out by Bothe and Gentner,[20] it is definitely possible that H^4 is stable. Li^4 is, of course, less stable because of the Coulomb repulsion between its three protons. If it is stable, Li^4 would be formed when He^3 captures a proton and would thus play an important role in stars (cf. §5). The only possible estimate of the stability seems to be a comparison

---

[17] H. A. Bethe, Rev. Mod. Phys. 9, 167 (1937).
[18] T. W. Bonner, Phys. Rev. 52, 685 (1938).
[19] H. A. Bethe, Phys. Rev. 53, 313 (1938).
[20] W. Bothe and W. Gentner, Naturwiss. 24, 17 (1936).

베테 논문 영문본

of Li^{4}Li^5 to the analogous pair B^{8}B^9. Reasonable estimates[21] make B^9 just unstable (0.3–0.7 mMU), while B^8 comes out at the limit of stability (binding energy between -0.4 and $+0.4$ mMU), i.e., B^8 about 0.3 to 0.7 mMU more stable than B^9. Li^5 (see below) is unstable by 1.4–1.8 mMU; if one assumes the same difference, Li^4 is found to be unstable by ~ 1 mMU. However, this argument is very uncertain and the possibility of a stable Li^4 cannot be excluded at present. H^4 would, from a similar argument, turn out stable by 0.6 mMU.

He^5 and Li^5

The instability of He^5 is shown directly by the experiments of Williams, Shepherd and Haxby[22] on the reaction

$$\text{Li}^7 + \text{H}^2 = \text{He}^5 + \text{He}^4. \qquad (15)$$

From a measurement of the range of the α-particles, the mass of He^5 is found (reference 23, Table 73, p. 373) to be 5.0137 whereas the combined mass of an α-particle plus a neutron would be only $4.003\,86 + 1.008\,93 = 5.012\,8$. Thus He^5 is unstable by 0.9 mMU (milli-mass-units) which is far outside the experimental error (about 0.1–0.2 mMU). It might be argued that the α-particle group observed in reaction (15) might correspond to an excited state of He^5. However, this is extremely improbable because a nucleus of such a simple structure as He^5 (α-particle plus neutron) should not have any low-lying excited levels.[24] (This holds both in the α-particle and the Hartree model of nuclear structure.) Moreover, it would be difficult to explain why the α-particle group corresponding to the ground state of He^5 should not have been observed.

The conclusion that the mass of 5.0137 found by W. S. and H. really corresponds to the ground state of He^5 is supported by considerations of mass defects. In fact, the instability of He^5 was

first predicted by Atkinson[25] on the basis of such considerations. Considering analogous nuclei, consisting of α-particles plus one neutron, we find that the last neutron is bound with a binding energy of 5.3 mMU in C^{13} and with only 1.8 mMU in Be^9. A binding energy of -0.9 mMU in He^5 fits very well into this series while a positive binding energy would not.

If He^5 is unstable, this is *a fortiori* true of Li^5 since the binding energies of these two nuclei should differ by just the Coulomb repulsion between proton and alpha in Li^5. This repulsion will be about 0.6–1 mMU, so that Li^5 is unstable by 1.4–1.8 mMU.

Thus all the nuclear evidence[25a] points to the nonexistence of both He^5 and Li^5. Even if there were no such evidence, astrophysical data themselves would force us to this conclusion, because at a temperature of $2 \cdot 10^7$ degrees (central temperature of sun) the energy production from the combination of He^4 and H forming Li^5 would be of the order of 10^{10} ergs/g sec. (cf. §4), as against an observed energy production of 2 ergs/g sec. Only the nonexistence of Li^5 prevents this enormous production of energy.

Be^6

This nucleus is certainly unstable, as can be shown by comparing it with the known nucleus He^6 from which it differs by the interchange of protons and neutrons. The Coulomb energy which is the only difference between the binding energies of the two nuclei can be calculated rather accurately.[21] The instability against disintegration into He$^4 + 2$H is between 1 and 2.6 mMU.

Be^7

This nucleus has been observed by Roberts, Heydenburg and Locher.[26] It decays with a

[21] H. A. Bethe, Phys. Rev. **54**, 436, 955 (1938).
[22] J. H. Williams, W. G. Shepherd and R. O. Haxby, Phys. Rev. **51**, 888 (1937).
[23] M. S. Livingston and H. A. Bethe, Rev. Mod. Phys. **9**, 247 (1937).
[24] With the possible exception of a doublet structure of the ground state, similar to Li^7. However, the doublet separation should probably be much smaller than in Li^7 because of the loose binding of He^5, and presumably both components of the doublet are already contained in the rather broad α-particle group observed by Williams, Shepherd and Haxby.

[25] R. d'E. Atkinson, Phys. Rev. **48**, 382 (1935).
[25a] *Note added in proof:*—Recently, F. Joliot and I. Zlotowski (J. de phys. et rad. **9**, 403 (1938)) reported the formation of stable He^5 from the reaction He$^4 + H^2 = $He^5 $+$H^1. The evidence is based upon the emission of singly charged particles of long range when heavy paraffin is bombarded by α-particles. However, the number of such particles observed was exceedingly small (only 6 out of a total of 126 tracks). Furthermore, the mass given for He^5 by Joliot and Zlotowski (5.0106) is irreconcilable with the stability (against neutron emission) of the well-known nucleus He^6.
[26] R. B. Roberts, N. P. Heydenburg and G. L. Locher, Phys. Rev. **53**, 1016 (1938).

세상에서 가장 쉬운 과학 수업 별의 물리학

TABLE IV. *Corrected and additional nuclear masses, and binding energies.*

NUCLEUS	MASS	BINDING ENERGY (mMU)	REFERENCE
n^1	1.008 93		19
He^3	3.016 99	5.87	18
H^4	4.025 4	0.6 $\pm$1	
He^4	4.003 86		29
Li^4	4.026 9	-1 $\pm$1	
He^5	5.013 7	-0.9 $\pm$0.2	23
Li^5	5.013 6	-1.6 $\pm$0.3	
Be^6	6.021 9	-1.8 $\pm$0.8	21
Be^7	7.019 28	5.7	26
Be^8	8.007 80	$-0.08$$\pm$0.04	28
B^8	8.027 4	0.0 $\pm$0.4	21
B^9	9.016 4	-0.5 $\pm$0.2	21
C^{10}	10.020 2	3.8	21
N^{12}	12.022 5 -24 3	0.0 $\pm$0.9	21
N^{13}	13.010 08	2.03	19
O^{14}	14.013 1	5.1	21

half-life of 43 days (mean life 60 days) and probably only captures K electrons. Calculations of the Coulomb energy,[21] on the other hand, would make positron emission just possible (positron energy $\sim$0.1 mMU). As a compromise, we assume that the mass of Be^7 is just equal to Li^7 plus two electrons, i.e., 7.019 28.

Be^8

The instability of Be^8 against disintegration into two α-particles has been definitely established by Paneth and Glückauf[27] who have shown that the Be^8 formed in the photoelectric disintegration of Be^9 disintegrates into 2 He^4. Kirchner and Neuert[28] have confirmed this conclusion by investigating the products of the disintegration $B^{11}+H=Be^8+He^4$. They found that frequently two α-particles enter the detecting apparatus simultaneously, with a small angle (less than 50°) between their respective directions of flight; this is just what should be expected if the Be^8 formed breaks up into two α-particles on its way to the detector. From the average angle between the two alphas, the disintegration energy of Be^8 (difference Be^8-2He^4) was estimated as between 40 and 120 kev.[28a]

---

[27] F. A. Paneth and E. Glückauf, Nature **139**, 712 (1937).
[28] F. Kirchner and H. Neuert, Naturwiss. **25**, 48 (1937).
[28a] *Note added in proof:*—These conclusions are compatible with the new measurements of S. K. Allison, E. R. Graves, L. S. Skaggs and N. M. Smith, Jr. (Phys. Rev. **55**, 107 (1939)) on the reaction energy of $Be^9+H = Be^8+H^2$.
[29] K. T. Bainbridge, Phys. Rev. **53**, 922(A) (1938).

B^8

The existence is doubtful; calculation[21] by comparison with its isobar Li^8 gives a binding energy between -0.4 and $+0.4$ mMU. This nucleus is not very important for astrophysics.

B^9

B^9 is almost certainly unstable, as can be shown by calculating[21] the difference in binding energy (Coulomb energy) between it and its isobar Be^9. The theoretical instability is between 0.3 and 0.7 mMU, 0.3 being almost certainly a lower limit. However, in view of the smallness of this instability, we shall at least discuss what would happen if B^9 were stable (§6). It will turn out that this would make almost no difference at "ordinary" temperatures ($2 \cdot 10^7$ degrees) and not much even at higher ones (10^8 degrees). For these calculations, we shall assume B^9 to be stable with 0.2 mMU which seems generous.

C^{10}

C^{10} is stable with 4 mMU against Be^8+2H.

N^{12}

N^{12} is doubtful, mainly because the binding energy of its isobar, B^{12}, is known only very inaccurately (between 2 and 3.3 mMU). Assuming 2 mMU, N^{12} would probably be instable, with 3.3 stable.

Table IV summarizes the binding energies of doubtful nuclei, and also gives some nuclear masses supplementary to and correcting those given in reference 23, p. 373.

§4. REACTION RATES AT $2 \cdot 10^7$ DEGREES

We are now prepared to actually calculate the rate of nuclear reactions under the conditions prevailing in stars. We choose a temperature of twenty million degrees, close to the temperature at the center of the sun. In order to have a figure independent of density and chemical composition, we calculate (cf. 7)

$$P = (m_2/x_2)p/\rho x_1, \qquad (16)$$

$P\rho x_1$ gives the probability (per second) that a given nucleus of kind 2 undergoes a reaction with any nucleus of kind 1. If there are no other reactions destroying or producing nuclei of kind 2, $1/P\rho x_1$ will be the mean life of nuclei of kind 2 in the star.

베테 논문 영문본

TABLE V. *Probability of nuclear reactions at $2 \cdot 10^7$ degrees.*[**]

REACTION	Q (mMU)	Γ (EV)	τ	P (SEC.⁻¹)	LIFE, FOR $\rho x_1 = 30$
$H + H = H^2 + \epsilon^+$	1.53	Ref. 16	12.5	$8.5 \cdot 10^{-21}$	$1.2 \cdot 10^{11}$ yr.
$H^2 + H = He^3$	5.9	$1\,E$	13.8	$1.3 \cdot 10^{-2}$	2 sec.
$H^3 + H = He^4$	21.3	$10\,E$	14.3	$1.7 \cdot 10^{-1}$	0.2 sec.
$He^3 + H = Li^{4*}$	(0.5)	$0.02\,D$	22.7	$3 \cdot 10^{-7}$	1 day
$He^4 + H = Li^{5*}$	(0.2)	$0.005\,D$	23.2	$6 \cdot 10^{-8}$	6 days
$Li^6 + H = He^4 + He^3$	4.1	$5 \cdot 10^5\,X$	31.1	$7 \cdot 10^{-3}$	5 sec.
$Li^7 + H = 2\,He^4$	18.6	$4 \cdot 10^4 X$	31.3	$6 \cdot 10^{-4}$	1 min.
$Be^7 + H = B^8?$	(0.5)	$0.02\,D$	38.1	$6 \cdot 10^{-13}$	2000 yr.
$Be^9 + H = Li^6 + He^4$	2.4	$10^6\,X$	38.1	$4 \cdot 10^{-5}$	15 min.
$B^9 + H = C^{10*}$	3.5	$2\,D$	44.6	$2 \cdot 10^{-13}$	5000 yr.
$B^{10} + H = C^{11}$	9.2	$10\,D$	44.6	10^{-12}	1000 yr.
$B^{11} + H = 3\,He^4$	9.4	$10^6\,E$	44.6	$1.2 \cdot 10^{-7}$	3 days
$C^{11} + H = N^{12}$	(0.4)	$0.02\,D$	50.6	10^{-17}	10^8 yr.
$C^{12} + H = N^{13}$	2.0	$0.6\,X$	50.6	$4 \cdot 10^{-16}$	$2.5 \cdot 10^6$ yr.
$C^{13} + H = N^{14}$	8.2	$30\,X$	50.6	$2 \cdot 10^{-14}$	$5 \cdot 10^4$ yr.
$N^{14} + H = O^{15}$	7.8	$5\,D$	56.3	$2 \cdot 10^{-17}$	$5 \cdot 10^7$ yr.
$N^{15} + H = C^{12} + He^4$	5.2	$10^6\,E$	56.3	$5 \cdot 10^{-13}$	2000 yr.
$O^{16} + H = F^{17}$	0.5	$0.02\,D$	61.6	$8 \cdot 10^{-22}$	10^{12} yr.
$F^{19} + H = O^{16} + He^4$	8.8	$10^6\,E$	66.9	$4 \cdot 10^{-17}$	$3 \cdot 10^7$ yr.
$Ne^{22} + H = Na^{23}$	10.7	$10\,D$	71.7	$5 \cdot 10^{-23}$	$2 \cdot 10^{13}$ yr.
$Mg^{26} + H = Al^{27}$	8.0	$10\,D$	81.3	10^{-26}	10^{17} yr.
$Si^{30} + H = P^{31}$	7.0	$10\,D$	90.4	$4 \cdot 10^{-30}$	$3 \cdot 10^{20}$ yr.
$Cl^{37} + H = A^{38}$	12.0	$10\,D$	103.1	$5 \cdot 10^{-35}$	$2 \cdot 10^{25}$ yr.
$H^2 + H^2 = He^3 + n$	3.5	$3 \cdot 10^5\,X$	15.7	10^3	
$Be^7 + H^2 = B^{9*}$	18.5	$10\,D'$	45.9	$2 \cdot 10^{-13}$	
$Be^7 + H^3 = B^9 + n^*$	11.9	$10^6\,E$	50.7	$2 \cdot 10^{-10}$	
$Be^7 + He^3 = C^{10}$	16.2	$1\,D'$	80.5	$3 \cdot 10^{-28}$	
$H^2 + He^4 = Li^6$	1.7	$4 \cdot 10^{-3}\,Q$	27.5	$3 \cdot 10^{-10}$	
$He^3 + He^4 = Be^7$	1.6	$0.02\,D'$	47.3	$3 \cdot 10^{-17}$	$3 \cdot 10^7$ yr.
$He^4 + He^4 = Be^{8*}$	(0.05)	$5 \cdot 10^{-9}\,Q$	50.0	10^{-24}	
$Li^7 + He^4 = B^{11}$	9.1	$1\,D'$	71.0	$2.5 \cdot 10^{-24}$	
$Be^7 + He^4 = C^{11}$	8.0	$1\,D'$	86	$3 \cdot 10^{-30}$	$3 \cdot 10^{20}$ yr.
$C^{12} + He^4 = O^{16}$	7.8	$1\,Q'$	119	$7 \cdot 10^{-43}$	

[**] The letters in the column giving the level width mean: X =experimental value; D =calculated for dipole radiation, from Eq. (12); D' = dipole radiation with small specific charge, 1/4 to 1/20 of Eq. (12); Q =quadrupole radiation, Eq. (12a); and E =estimate.
[*] These reactions are not believed to occur since their product or one of the reactants is unstable. They are listed merely for the sake of discussion.

Table V gives the results of the calculations, based on Eqs. (7) to (9). In the first column, the nuclear reactions are listed. All reactions which seemed of importance in the interior of stars were considered; in addition, some reactions with heavier elements (O^{16} to Cl^{37}) were included in order to show the manner in which the reaction rate decreases. Moreover, seven reactions were listed in spite of the fact that their products or reactants are believed to be (§3) unstable (starred) or doubtful (question mark); these reactions are included in order to discuss the consequences if they did occur.

The second column gives the energy evolution Q in the reaction, calculated from the masses (reference 23, Table LXXIII, and this paper, Table IV). In the third column, the width Γ determining the reaction rate (cf. §2) is tabulated. Wherever possible, this was taken from experiments (Tables II and III) or from the "empirical formulae" (12), (12a) for the radiation width. For the radiative combination of two nuclei of equal specific charge ($H^2 + He^4$, $He^4 + He^4$, $C^{12} + He^4$) quadrupole radiation was assumed, otherwise dipole radiation.[30] For almost equal specific charge (e.g., $Be^7 + He^4$), the dipole formula with an appropriate reduction was used. In some instances, the width was estimated by analogy (e.g., $N^{15} + H = C^{12} + He^4$) or from approximate theoretical calculations ($H^2 + H = He^3$).[31] The way in which Γ was obtained was indicated by a letter in each instance.

The fourth column contains τ, as calculated from (8), the fifth P from (16). The wide variation of P is evident, also the smallness of P for α-particle as compared with proton reactions.

[30] If the combined initial nuclei and the final nucleus have the same parity (as may be the case, e.g., for $O^{16} + H = F^{17}$), it is still possible to have a dipole transition if only the incident particle has orbital momentum one. This does not materially affect its penetrability if $R > a$ (cf. (4), (5)) which is true in every case where the parities are expected to be the same.

[31] L. I. Schiff, Phys. Rev. 52, 242 (1937).

E.g., the reaction between He^4 and a nucleus as light as Be is as improbable as between a proton and Si. This arises, of course, from the greater charge and mass of the α-particle both of which factors reduce its penetrability. The reaction $He^4 + He^4 = Be^8$ has an exceedingly small probability because of the small frequency and the quadrupole character of the emitted γ-rays. Thus this reaction would not be important even if Be^8 were stable. On the other hand, the reaction $He^4 + H = Li^5$ would be extremely probable if Li^5 existed. The helium in the sun would be "burnt up" completely in about six days, even if rather unfavorable assumptions are made about the probability of the reaction. Similarly, if the energy evolution per process is 0.2 mMU $= 3 \cdot 10^{-7}$ ergs, the energy produced per gram of the star would be

$$(6 \cdot 10^{23}/4)\rho x_H x_{He} \cdot 3 \cdot 10^{-7} \cdot 6 \cdot 10^{-8}.$$

With $\rho = 80$, $x_H = 0.35$ and $x_{He} = 0.1$, this would give about 10^{10} ergs/g sec. as against 2 ergs/g sec. observed. This is a very strong additional argument against the existence of Li^5.

In the last column of Table V, the mean life is calculated for the various nuclei reacting with protons, by assuming a density $\rho = 80$ and hydrogen content $x_1 = 35$ percent, which correspond to the values at the center of the sun.[1] It is seen that, with the exception of H, the lifetimes of all nuclei up to boron are quite short, ranging from a fraction of a second for H^3 to 1000 years for B^{10}. (The life of B^{10} may actually be slightly shorter because of the reaction $B^{10} + H = Be^7 + He^4$. See §6.) Of the two lives longer than 1000 years listed, one refers to B^9 which probably does not exist (§3), the other to Be^7 which decays by positron emission with a half-life of 43 days.[26] We must conclude that *all the nuclei between H and C, notably H^2, H^3, Li^6, Li^7, Be^9, B^{10}, B^{11}, can exist in the interior of stars only to the extent to which they are continuously re-formed by nuclear reactions.* This conclusion does not apply to He^4 because Li^5 does not exist. To He^3 it probably applies whether Li^4 exists or not, because He^3 will also be destroyed by combination with He^4 into Be^7, although with a considerably longer period ($3 \cdot 10^7$ years instead of the 1 day for the reaction giving Li^4).

The actual lifetime of carbon and nitrogen is much longer than it would appear from the table because these nuclei are reproduced by the nuclear reactions themselves (§7). This makes their actual lifetime of the order of 10^{12} (or even 10^{20}, cf. §7) years, i.e., long compared with the age of the universe ($\sim 2 \cdot 10^9$ years). Protons, and all nuclei heavier than nitrogen, also have lives long compared with astronomical times.

§5. THE REACTIONS FOLLOWING PROTON COMBINATION

In the last section, it has been shown that all elements lighter than carbon, with the exception of H^1 and He^4, have an exceedingly short life in the interior of stars. Such elements can therefore only be present to the extent to which they are continuously produced in nuclear reactions from elements of longer life. This is in accord with the small abundance of all these elements both in stars and on earth.

Of the two more stable nuclei, He^4 is too inert to play an important rôle. It combines neither with a proton nor with another α-particle since the product would in both cases be an unstable nucleus. The only way in which He^4 can react at all, is by triple collisions. These will be discussed in the next section and will be shown to be very rare, as is to be expected.

As the only primary reaction between elements lighter than carbon, there remains therefore the reaction between two protons,

$$H^1 + H^1 = H^2 + \epsilon^+. \tag{1}$$

According to Critchfield and Bethe,[16] this process gives an energy evolution of 2.2 ergs/g sec. under "standard stellar conditions" ($2 \cdot 10^7$ degrees, $\rho = 80$, hydrogen content 35 percent). The reaction rate under these conditions is (cf. Table V) $2.5 \cdot 10^{-19}$ sec.$^{-1}$, corresponding to a mean life of $1.2 \cdot 10^{11}$ years for the hydrogen in the sun. This lifetime is about 70 times the age of the universe as obtained from the red shift of nebulae.

According to the foregoing, any building up of elements out of hydrogen will have to start with reaction (1). The deuteron will capture another proton,

$$H^2 + H^1 = He^3. \tag{17}$$

This reaction follows almost instantaneously upon (1), with a delay of only 2 sec. (Table V). There is, therefore, always statistical equilibrium between protons and deuterons, the concentrations (by number of atoms) being in the ratio of the respective lifetimes. This makes the concentration of deuterons (by weight) equal to

$$x(H^2) = x(H^1) \cdot \frac{2 \cdot 8.5 \cdot 10^{-21}}{1.3 \cdot 10^{-2}}$$

$$= 1.3 \cdot 10^{-18} x(H^1) \quad (18)$$

(cf. Table V). The relative probability of the reaction

$$H^2 + H^2 = He^3 + n^1 \quad (19)$$

as compared with (17), is then

$$p_n = \frac{1}{4} \cdot \frac{P(H^2 + H^2 = He^3 + n^1)}{P(H^2 + H = He^3)} \cdot \frac{x(H^2)}{x(H^1)}$$

$$= \frac{10^3 \cdot 1.3 \cdot 10^{-18}}{4 \cdot 1.3 \cdot 10^{-2}} = 2 \cdot 10^{-14}. \quad (19a)$$

(One factor $\frac{1}{2}$ comes from the fact that in (19) two nuclei of the same kind interact; another is the atomic weight of H^2.) Thus one neutron is produced for about $5 \cdot 10^{13}$ proton combinations.

The further development of the He^3 produced according to (17) depends on the question of the stability of Li^4 and of the relative stability of H^3 and He^3.

Assumption A: Li⁴ stable

In this case, the He^3 will capture another proton, viz.:

$$He^3 + H^1 = Li^4. \quad (20)$$

With the assumptions made in Table V, the mean life of He^3 would be 1 day. The Li^4 would then emit a positron:

$$Li^4 = He^4 + \epsilon^+. \quad (20a)$$

With an assumed stability of Li^4 of 0.5 mMU compared with $He^3 + H^1$, the maximum energy of the positrons in (20a) would be 20.8 mMU = 19.4 Mev (including rest mass) which would be by far the highest β-ray energy known. The lifetime of Li^4 may accordingly be expected to be a small fraction of a second (half-life = 1/500 sec. for an allowed transition in the Fermi theory).

The most important consequence of the stability of Li^4 would be that only a fraction of the mass difference between four protons and an α-particle would appear as usable energy. For in the β-emission (20a) the larger part of the energy is, on the average, given to the neutrino which will in general leave the star without giving up any of its energy

(see below). According to the Konopinski-Uhlenbeck theory, which is in good agreement with the observed energy distribution in β-spectra, the neutrino receives on the average 5/8 of the total available energy if the latter is large. In our case, this would be 13.0 mMU. Adding 0.2 mMU for the neutrino emitted in process (1), we find that altogether 13.2 mMU energy is lost to neutrinos, of a total of 28.7 mMU developed in the formation of an α-particle out of four protons and two electrons. Thus the observable energy evolution is only 15.5 mMU, i.e., 54 percent of the total. Therefore, if Li^4 is stable, process (1) would give only 1.2 ergs/g sec. instead of 2.2 (under "standard" conditions).[16]

The neutrinos emitted will have some chance of producing neutrons in the outer layers of the star. It seems reasonable to assume that a neutrino has no other interaction with matter than that implied in the β-theory. Then a free neutrino (ν) will cause only "reverse β-processes"[32] of which the simplest and most probable is

$$H + \nu = n^1 + \epsilon^+. \quad (21)$$

This process is endoergic with 1.9 mMU and is therefore caused only by fast neutrinos such as those from Li^4. The cross section is according to the Fermi theory

$$\sigma = \pi^2 (h/mc)^3 \cdot 0.9 \cdot 10^{-4} c^{-1} W (W^2 - 1)^{\frac{1}{2}}$$
$$= 1.7 \cdot 10^{-45} W (W^2 - 1)^{\frac{1}{2}} \text{ cm}^2, \quad (22)$$

where W is the energy of the emitted positron, including rest mass, in units of mc^2. In reaction (21), this is the neutrino energy minus 1.35 mMU. For the Li^4 neutrinos, the average cross section comes out to be

$$\sigma_{Av} = 2.5 \cdot 10^{-43} \text{ cm}^2 \quad (22a)$$

per proton, and the probability of process (21) for a neutrino starting from the center of the star

$$p = 6 \cdot 10^{23} \cdot \sigma_{Av} x_H \int_0^R \rho dr = 1.5 \cdot 10^{-19} x_H \int_0^R \rho dr, \quad (22b)$$

where $\rho(r)$ is the density (in g/cm³) at the distance r from the center of the star. For the sun, $p = 1.6 \cdot 10^{-7}$. This means that $1.6 \cdot 10^{-7}$ of the neutrinos emitted will cause reaction (21) before leaving the sun, and that the number of neutrons formed is $1.6 \cdot 10^{-7}$ times the number of proton combinations (1).

A further consequence of (20, 20a) would be that ordinarily no nuclei heavier than 4 mass units are formed at all, even as intermediate products. Such nuclei would only be produced in the rare cases when H^2 or He^3 capture an α-particle rather than a proton, according to the reactions

$$H^2 + He^4 = Li^6 \quad (23)$$
and
$$He^3 + He^4 = Be^7. \quad (24)$$

Under the favorable assumption that the concentration of He^4 is the same as of H^1 (by weight), the fraction of H^2 forming Li^6 is (cf. Table V)

$$p(Li^6) = 3 \cdot 10^{-10} / 1.3 \cdot 10^{-2} = 2 \cdot 10^{-8} \quad (23a)$$

the fraction of He^3 giving Be^7 is

$$p(Be^7) = 3 \cdot 10^{-17} / 3 \cdot 10^{-7} = 10^{-10}. \quad (24a)$$

[32] H. A. Bethe and R. Peierls, Nature 133, 689 (1934).

세상에서 가장 쉬운 과학 수업 별의 물리학

Most of the Li^6 will give rise to the well-known reaction

$$Li^6 + H = He^3 + He^4 \qquad (23b)$$

and most of the Be^7 will go over into Li^7 which in turn reacts with a proton to give two α-particles. Only occasionally, Li^6, Be^7 or Li^7 will capture an α-particle and thus form heavier nuclei. It can be shown (cf. assumption B) that Li^7 is the most efficient nucleus in this respect. Therefore, the amount of heavier elements formed is determined by Be^7, the mother substance of Li^7, and is thus 10^{-10} times the amount formed with assumption B.

Assumption B: Li^4 unstable, He^3 more stable than H^3

This assumption seems to be the most likely according to available evidence. The *only* reaction which the He^3 can undergo, is then (24), i.e., each proton combination leads to the formation of a Be^7 nucleus. The most probable mode of decay of this nucleus is by electron capture, leading to Li^7. The lifetime of Be^7 (half-life) is 43 days[26] in the complete atom, and 10 months at the center of the sun (cf. 14, 14a). This makes the mean life $= 14$ months and the reaction rate $2.8 \cdot 10^{-8}$ sec.$^{-1}$. The capture of a proton by Be^7 would, even if the product B^8 is stable, be 2000 times slower (Table V). Each electron capture by Be^7 is accompanied by the emission of a neutrino of energy $\sim 2mc^2 = 1.1$ mMU (when Li^7 is left in its excited state, which happens rather rarely, the neutrino receives only 0.6 mMU). The total energy lost to neutrinos (including process 1) will therefore be very small in this case (~ 1.3 mMU per α-particle formed, i.e., $4\frac{1}{2}$ percent of the total energy evolution) and practically the full mass energy will be transformed into heat radiation. The Li^7 formed by electron capture of Be^7 will cause the well-known reaction

$$Li^7 + H = 2 \, He^4 \qquad (25)$$

and have (Table V) a mean life of only 1 minute at $2 \cdot 10^7$ degrees.

The reaction chain described leads, as in the case of assumption A, to the building up of one α-particle out of four protons and two electrons, for each process (1). No nuclei heavier than He^4 are formed permanently. Such nuclei can be produced only by branch reactions alternative to the main chain described. These will be discussed in the following.

a. Reactions with protons.—When Li^7 reacts with slow protons, the result is not always two α-particles, but, in one case out of about[33] 5000, radiative capture, giving Be^8. However, Be^8 will disintegrate again into two α-particles (§3), and during its life of about 10^{-13} sec., the probability of its reacting with another particle (e.g., capture of another proton) is exceedingly small ($\sim 10^{-24}$). Similarly, Be^7 will, in one out of about 2000 cases (see above) capture a proton and form B^8 if that nucleus exists. However, B^8

[33] It was assumed that radiative capture takes place only through the resonance level at 440 kv proton energy. The proton width of this level was taken as 11 kv, the radiation width as 40 ev.

will again go over into Be^8, by positron emission, and two alphas will again be the final result.

At this place, obviously, stability of Be^8 would increase the yield of heavy nuclei. Then one stable Be^8 would be formed for 5000 proton combinations; and, if B^9 is also assumed to be stable, every Be^8 goes over into B^9. Since about one out of $3 \cdot 10^8$ B^9 gives a C^{12} (§6), the number of heavy nuclei (C^{12}) formed would be $\sim 10^{-12}$ per α-particle. This would be the highest yield obtainable. However, Be^8 is known to be unstable (§3).

b. Reactions with α-particles.—The only abundant light nucleus other than the proton is He^4. The only reaction possible between an α-particle and Li^7 or Be^7 is radiative capture, viz.

$$Li^7 + He^4 = B^{11}, \qquad (26)$$
$$Be^7 + He^4 = C^{11}. \qquad (26a)$$

The probability of formation of B^{11} and C^{11} is (Table V)

$$p(B^{11}) = \frac{P(Li^7 + He^4)}{P(Li^7 + H)} = \frac{2.5 \cdot 10^{-24}}{6 \cdot 10^{-4}} = 4 \cdot 10^{-21}, \qquad (26b)$$

$$p(C^{11}) = \frac{P(Be^7 + He^4)}{P(Be^7 + \epsilon = Li^7)} = \frac{14 \text{ months}}{3 \cdot 10^{20} \text{ yr.}} = 4 \cdot 10^{-21}. \qquad (26c)$$

Thus the formation of B^{11} is about as probable as that of C^{11}; the effect of the lower potential barrier of Li^7 for α-particles is compensated by its shorter life. The C^{11} will, of course, also give B^{11} by positron emission.

The B^{11} will react with protons in two ways, viz.

$$B^{11} + H^1 = 3 \, He^4, \qquad (27)$$
$$B^{11} + H^1 = C^{12}. \qquad (27a)$$

The branching ratio is about $10^4 : 1$ in favor of (27) (calculated from experimental data). Thus there will be one C^{12} nucleus formed for about 10^{24} α-particles. The building up of heavier nuclei, even in this most favorable case, is therefore exceedingly improbable.

c. Reactions with He^3.—Since He^3 has a rather long life $3 \cdot 10^7$ years, Table V) and penetrates more easily through the potential barrier than the heavier He^4, it may be considered as an alternative possibility. However, the probability of formation of C^{10} from $Be^7 + He^3$ is only 100 times greater than that of C^{11} from $Be^7 + He^4$ (Table V) if the concentrations of He^3 and He^4 were equal. Actually, that of He^3 is only about $3 \cdot 10^{-4}$ (life of He^3 divided by life of protons) so that this process is 1/30 as probable as (26a). For $Li^7 + He^3$, the situation is even less favorable.

d. Reactions with H^2.—Deuteron capture by Be^7 would lead to B^9 whose existence is very doubtful. The probability per second would be (cf. Table V and Eq. (18))

$$2 \cdot 10^{-13} \rho x(H^2) = 2 \cdot 10^{-13} \cdot 30 \cdot 1.3 \cdot 10^{-18} = 10^{-29}, \qquad (28)$$

which is only 1/10 of the probability of (26a) (Table V). Moreover, most of the B^9 formed reverts to He^4 (§6) so that the contribution of this process is negligible.

e. Reactions with He^4 in statu nascendi.—The process (25) produces continuously fast α-particles which need not penetrate through potential barriers. These α-particles have a range of 8 cm each in standard air, corresponding to 16 cm $= 0.02$ g/cm^2 for both. In stars, with their large hydrogen content, a somewhat smaller figure must be used since hydrogen has a greater stopping power per gram; we take 0.01 g/cm². The cross section for fast particles is about

$$\sigma = \pi R^2 \frac{\Gamma}{\hbar^2/MR^2}. \qquad (29)$$

With $\Gamma = 1$ volt (Table V) and $R = 3.6 \cdot 10^{-13}$ cm (Eq. (10)), this gives $\sigma = 1.3 \cdot 10^{-31}$ cm². The number of Be⁷ atoms per gram is

$$6 \cdot 10^{23} x(\text{Be}^7) = 6 \cdot 10^{23} \cdot \frac{14 \text{ months}}{1.2 \cdot 10^{11} \text{ yr}} x(\text{H}) = 2 \cdot 10^{12} \qquad (29a)$$

with $x(\text{H}) = 0.35$. This gives for the number of processes (26a) per proton combination:

$$0.01 \cdot 2 \cdot 10^{12} \cdot 1.3 \cdot 10^{-31} = 2.5 \cdot 10^{-21}, \qquad (29b)$$

which is about the same as the formation of C¹¹ or B¹¹ by capture of *slow* alphas (cf. 26b, c).

Returning to the main reaction chain in the case of our assumption B, we note that the formation of Be⁷ (Eq. (24)) is a very slow reaction, requiring $3 \cdot 10^7$ years at "standard" conditions ($2 \cdot 10^7$ degrees). At lower temperatures, the reaction will be still slower and, finally, it will take longer than the past life of the universe ($\sim 2 \cdot 10^9$ years). In this case, the amount of He³ present will be much smaller than its equilibrium value (provided there was no He³ "in the beginning") and the energy production due to reactions (24), (25) will be reduced accordingly. Ultimately, at very low temperatures ($< 12 \cdot 10^6$ degrees), the reaction H+H will lead only to He³, and will therefore give an energy production of only 7.2 mMU, i.e., only one-quarter of the high temperature value, 27.4 mMU.

Assumption C: H³ more stable than He³

In this case, He³ will be able to capture an electron,

$$\text{He}^3 + \epsilon^- = \text{H}^3. \qquad (30)$$

Under the assumption that a difference in mass of 0.1 electron mass exists between He³ and H³, the probability of (30) is, according to (14a),

$$p(\text{H}^3) = 1.5 \cdot 10^{-11} \text{ sec.}^{-1} \qquad (30a)$$

for a density $\rho = 80$ and 35 percent hydrogen content. This corresponds to a mean life of ~ 2000 years. The electron capture is therefore about 10^4 times more probable than the formation of Be⁷ according to (24). This ratio will be reversed at temperatures $> 4 \cdot 10^7$ since (30) is independent of T and the probability of (24) increases as T^{15}.

H³ will capture a proton and form He⁴,

$$\text{H}^3 + \text{H} = \text{He}^4,$$

with a mean life of about 0.2 seconds. This way of formation of He⁴ from reaction (1) is probably the most direct of all. As in B, practically no energy is lost to neutrinos.

The formation of heavier elements goes as in B, but now there is only one Be⁷ formed for 10^4 proton combination processes. This reduces the probability of formation of C¹² by another factor 10^4, to one C¹² in 10^{28} alphas.

The H³ itself does not contribute appreciably to the building up of elements. It is true that the reaction Be⁷+H³ = B⁹+n¹ is about 100 times as probable as Be⁷+H² = B⁹ (considering the shorter life of H³), and therefore 10 times as probable as (26a). However, most of the B⁹ reverts to He⁴ (cf. §6) so that (26) and (26a) remain the most efficient processes for building up C¹².

Summarizing, we find that the formation of nuclei heavier than He⁴ can occur only in negligible amounts. One C¹² in 10^{24} α-particles and one neutron in 10^{14} α-particles are the yields when Li⁴ is unstable, one C¹² in 10^{34} alphas and one neutron in 10^7 alphas when Li⁴ is stable. The reason for the small probability of formation of C¹² is twofold: First, any nonradioactive nucleus between He and C, i.e., Li⁶, ⁷, Be⁹, B¹⁰, ¹¹, reacting with protons will give α-particle emission rather than radiative capture so that a disintegration takes place rather than a building up. This will no longer be the case for heavier nuclei so that for these a building up is actually possible. Second, the instability of Be⁸ causes a gap in the list of stable nuclei which is the harder to bridge because Be⁸ is very easily formed in nuclear reactions (small mass excess). On the other hand, the instability of He⁵ and Li⁵ is of no influence because Be⁷ and Li⁷ are stages in the ordinary chain of nuclear reactions.

§6. TRIPLE COLLISIONS OF ALPHA-PARTICLES

In the preceding section, we have shown that collisions with protons alone lead practically always to the formation of α-particles. In order that heavier nuclei be formed, use must therefore certainly be made of the α-particles themselves. However, collisions of an α-particle with one other particle, proton or alpha, do not lead to stable nuclei. Therefore we must assume triple collisions, of which three types are conceivable:

$$\text{He}^4 + 2\text{H} = \text{Be}^6, \qquad (31)$$

$$2\text{He}^4 + \text{H} = \text{B}^9, \qquad (32)$$

$$3\text{He}^4 = \text{C}^{12}. \qquad (33)$$

The first of these reactions leads to a nucleus which is certainly unstable (Be⁶). Even if it were stable, it would not offer any advantages over Be⁷ which is formed as a consequence of the proton combination (1). The second reaction leads to B⁹ which is probably also unstable. However, since this is not absolutely certain, we

[34] G. Breit and E. Wigner, Phys. Rev. **49**, 519 (1936).

세상에서 가장 쉬운 과학 수업 별의 물리학

shall discuss this process in the following. The last process leads directly to C^{12}, but since it involves a rather large potential barrier for the last α-particle, it is very improbable at $2 \cdot 10^7$ degrees (see below).

The formation of B^9

The probability of this process is enhanced by the well-known resonance level of Be^8, which corresponds to a kinetic energy of about $E = 50$ kev of two α-particles. The formation of B^9 occurs in two stages,

$$2 He^4 = Be^8, \quad Be^8 + H = B^9 \qquad (32a)$$

with a time interval of about 10^{-13} sec. (life of Be^8). The process can be treated with the usual formalism for resonance disintegrations, the compound nucleus being Be^8. This nucleus can "disintegrate" in two ways, (a) into two α-particles, (b) with proton capture. We denote the respective widths of the Be^8 level by Γ_α and Γ_H; the latter is given by the ordinary theory of thermonuclear processes, i.e.,

$$\Gamma_H = h p m_2 / x_2, \qquad (34)$$

where p is given by (4), and the subscripts 1 and 2 denote H^1 and Be^8, respectively.

The cross section of the resonance disintegration becomes then

$$\sigma = \pi \lambda^2 \frac{\Gamma_\alpha \Gamma_H}{(E - E_r)^2 + \frac{1}{4}(\Gamma_\alpha + \Gamma_H)^2}. \qquad (35)$$

E_r is the resonance energy. Γ_α is much larger than Γ_H (corresponding to about 10^{13} and 10^{-11} sec.$^{-1}$, respectively) but very small compared with E_r (about 10^{-2} against $5 \cdot 10^4$ volts). The resonance is thus very sharp, and, integrating over the energy, we obtain for the total number of processes per cm^3 per sec. simply

$$\rho p = B(E_r) v_r \pi \lambda_r^2 2\pi \Gamma_H. \qquad (36)$$

Here $B(E)dE$ is the number of pairs of α-particles with relative kinetic energy between E and $E + dE$ per cm^3, viz.

$$B(E) = \frac{1}{2}\left(\frac{\rho x_\alpha}{m_\alpha}\right)^2 \frac{2}{\pi^{\frac{1}{2}}} \frac{E^{\frac{1}{2}}}{(kT)^{\frac{3}{2}}} e^{-E/kT} \qquad (36a)$$

(x_α = concentration of He^4 by weight). Combining (36), (36a) and (34), (4), we find

$$p(B^9) = \frac{16\pi^{3/2}}{3^{5/2}} \frac{\rho^2 x_\alpha^2 x_H}{m_\alpha^{7/2} m_H} \frac{h^2 \Gamma_{rad}}{(kT)^{3/2}} a R^2 \tau^2$$
$$\times \exp\{4(2R/a)^{\frac{1}{2}} - \tau - E_r/kT\}, \qquad (37)$$

where Γ_{rad} is the radiation width for the process $Be^8 + H = B^9$. Numerically, (37) gives for the decay constant of hydrogen the value

$$p m_H / x_H = 1.00 \cdot 10^{-4}(\rho x_\alpha)^2 \Gamma \varphi T^{-1} e^{-11.7 E_r/T} \tau^2 e^{-\tau}, \qquad (37a)$$

where T is measured in millions of degrees, the resonance energy E_r of Be^8 in kilo-electron-volts and Γ in ev. The quantities Γ, φ and τ refer to the process $Be^8 + H = B^9$. If

there were no resonance level of Be^8, (37) would be replaced by

$$p(B^9) = \frac{8}{243} \frac{\rho^2 x_\alpha^2 x_H}{m_\alpha^2 m_H} \frac{\Gamma_{rad}}{\hbar} a R^2 a' R'^2 \tau^2 \tau'^2$$
$$\times \exp\{(32R/a)^{\frac{1}{2}} + (32R'/a')^{\frac{1}{2}} - \tau - \tau'\}, \qquad (38)$$

where the primed quantities refer to the reaction $2 He^4 = Be^8$, the unprimed ones to $Be^8 + H = B^9$. Numerically, (38) gives

$$p m_H / x_H = 2.0 \cdot 10^{-11} (\rho x_\alpha)^2 \Gamma \varphi \tau^2 e^{-\tau} \varphi' \tau'^2 e^{-\tau'}. \qquad (38a)$$

Assuming $T = 20$, $\rho = 80$, $x_\alpha = 0.25$, $\Gamma = 0.02$ electron-volts, we obtain for the probability of formation of B^9 per proton per second:

$$p m_H / x_H = 2 \cdot 10^{-26} \text{ for resonance, } E_r = 25 \text{ kev}$$
$$10^{-31} \text{ for resonance, } E_r = 50 \text{ kev}$$
$$4 \cdot 10^{-38} \text{ for resonance, } E_r = 75 \text{ kev}$$
$$2 \cdot 10^{-44} \text{ for resonance, } E_r = 100 \text{ kev}$$
$$5 \cdot 10^{-42} \text{ for nonresonance.}$$

The value 25 kev for the resonance level must probably be excluded on the basis of the experiments of Kirchner and Neuert.[28] But even for this low value of the resonance energy, the probability of formation of B^9 is only 10^{-6} times that of the proton combination $H + H = H^2 + \epsilon^+$ (Table V, $\rho x_1 = 30$). With $E_r = 50$ kev which seems a likely value, the ratio becomes $4 \cdot 10^{-13}$. On the other hand, the building up of B^9 (if this nucleus exists) would still be the most efficient process for obtaining heavier elements (see below).

Reactions of B^9

It can easily be seen[21] that B^9 cannot be positron-active but can only capture electrons if it exists at all. If B^9 is stable by 0.3 mMU, the energy evolution in electron capture would be just one electron mass. The decay constant of B^9 (for β-capture) is then, according to (14a), $1.5 \cdot 10^{-9}$ sec.$^{-1}$ ($\rho = 80$, $x_H = 0.35$) corresponding to a lifetime of about 20 years. On the other hand, the lifetime with respect to proton capture (Table V) is 5000 years. Therefore, ordinarily B^9 will go over into Be^9. This nucleus, in turn, will in general undergo one of the two well-known reactions:

$$Be^9 + H = Be^8 + H^2, \qquad (39)$$
$$Be^9 + H = Li^6 + He^4. \qquad (39a)$$

Only in one out of about 10^5 cases, B^{10} will be formed by radiative proton capture. Therefore the more efficient way for building up heavier elements will be the direct proton capture by B^9, leading to C^{10}, which occurs in one out of about 300 cases.

The C^{10} produced will go over into B^{10} by positron emission. B^{10} may react in either of the following ways:

$$B^{10} + H = C^{11}, \qquad (40)$$
$$B^{10} + H = Be^7 + He^4. \qquad (40a)$$

The reaction energy of (40a) is (cf. Table VII, §8) 1.2 mMU; the penetrability of the outgoing alphas about 1/40 (same table), therefore the probability of the particle reaction (40a) will be about 100 times that of the capture reaction (40).

The C^{11} from (40) will emit another positron. The resulting B^{11} reacts with protons as follows:

$$B^{11}+H=C^{12}, \qquad (41)$$
$$B^{11}+H=3\,He^4. \qquad (41a)$$

Both reactions are well-known experimentally. Reaction (41) has a resonance at 160 kev. From the width of this resonance and the experimental yields, the probability of (41) with low energy protons is about 1 in 10,000 (i.e., the same as for nonresonance). Altogether, about one B^9 in $3\cdot10^8$ will transform into C^{12}.

With a resonance energy of Be^8 of 50 kev, and $2\cdot10^7$ degrees, there will thus be about one C^{12} formed for 10^{21} α-particles if B^9 is stable. This is better than any other process but still negligibly small.

At higher temperatures, the formation of B^9 will become more probable and will, for $T>10^8$, exceed the probability of the proton combination. At these temperatures (actually already for $T>3\cdot10^7$) the B^9 will rather capture a proton (giving C^{10}). Even then, there remain the unfavorable branching ratios in reactions (40), (40a) and (41), (41a),[35] so that there will still be only one C^{12} formed in 10^6 alphas. Thus even with B^9 stable and granting the excessively high temperature, the amount of heavy nuclei formed is extremely small.

Direct formation of C^{12}

C^{12} may be formed directly in a collision between 3 α-particles. The calculation of the probability is exactly the same as for the formation of B^9. The nonresonance process gives about the same probability as a resonance of Be^8 at 50 kev. With $\rho=80$, $x_\alpha=\frac{1}{4}$, $\Gamma=0.1$ electron-volt, $T=2\cdot10^7$ degrees, the probability is 10^{-56} per α-particle, i.e., about 10^{-37} of the proton combination reaction (1). This gives an even smaller yield of C^{12} than the chains described in this and the preceding section. The process is strongly temperature-dependent, but it requires temperatures of $\sim10^9$ degrees to make it as probable as the proton combination (1).

The considerations of the last two sections show that there is no way in which nuclei heavier than helium can be produced permanently in the interior of stars under present conditions. We can therefore drop the discussion of the building up of elements entirely and can confine ourselves to the energy production which is, in fact, the only observable process in stars.

§7. The Carbon-Nitrogen Group

In contrast to lighter nuclei, C^{12} is not permanently destroyed when it reacts with protons;

---

[35] The reaction $C^{11}+H=N^{12}$ becomes more probable than $C^{11}=B^{11}+\epsilon^+$ only at $T>3\cdot10^8$ degrees. The branching ratio in (40), (40a) may perhaps be slightly more favorable because the effect of the potential barrier in (40a) may be stronger.

instead the following chain of reactions occurs

$$C^{12}+H^1=N^{13}, \qquad (42)$$
$$N^{13} \quad =C^{13}+\epsilon^+, \qquad (42a)$$
$$C^{13}+H^1=N^{14}, \qquad (42b)$$
$$N^{14}+H^1=O^{15}, \qquad (42c)$$
$$O^{15} \quad =N^{15}+\epsilon^+, \qquad (42d)$$
$$N^{15}+H^1=C^{12}+He^4. \qquad (42e)$$

Thus the C^{12} nucleus is reproduced. The reason is that the alternative reactions producing α-particles, *viz.*

$$C^{12}+H^1=B^9+He^4, \qquad (43)$$
$$C^{13}+H^1=B^{10}+He^4, \qquad (43a)$$
$$N^{14}+H^1=C^{11}+He^4, \qquad (43b)$$

are all strongly forbidden energetically (Table VII, §8). This in turn is due to the much greater stability of the nuclei in the carbon-nitrogen group as compared with the beryllium-boron group, and is in contrast to the reactions of Li, Be and B with protons which all lead to emission of α-particles.

The cyclical nature of the chain (42) means that practically no carbon will be consumed. Only in about 1 out of 10^4 cases, N^{15} will capture a proton rather than react according to (42e). In this case, O^{16} is formed:

$$N^{15}+H^1=O^{16}. \qquad (44)$$

However, even then the C^{12} is not permanently destroyed, because except in about one out of $5\cdot10^7$ cases, O^{16} will again return to C^{12} (cf. §8). Thus there is less than one C^{12} permanently consumed for 10^{12} protons. Since the concentration of carbon and nitrogen, according to the evidence from stellar spectra, is certainly greater than 10^{-12} this concentration does not change noticeably during the evolution of a star. *Carbon and nitrogen are true catalysts; what really takes place is the combination of four protons and two electrons into an α-particle.*

A given C^{12} nucleus will, at the center of the sun, capture a proton once in $2.5\cdot10^6$ years (Table V), a given N^{14} once in $5\cdot10^7$ years. These times are short compared with the age of the sun. Therefore the cycle (42) will have re-

세상에서 가장 쉬운 과학 수업 별의 물리학

TABLE VI. *Central temperatures necessary for giving observed energy production in sun, with various nuclear reactions.*

REACTION	T (MILLION DEGREES)
$H^2 + H = He^3$	0.36
$He^4 + H = Li^5$	2.1
$Li^7 + H = 2He^4$	2.2
$Be^9 + H = Li^6 + He^4$	3.3
$B^{10} + H = C^{11}$	9.2
$B^{11} + H = 3He^4$	5.5
$C^{12} + H = N^{13}$	15.5
$N^{14} + H = O^{15}$	18.3
$O^{16} + H = F^{17}$	32
$Ne^{22} + H = Na^{23}$	37

peated itself many times in the history of the sun, so that statistical equilibrium has been established between all the nuclei occurring in the cycle, viz. $C^{12}C^{13}N^{13}N^{14}N^{15}O^{15}$. In statistical equilibrium, the concentration of each nucleus is proportional to its lifetime. Therefore N^{14} should have the greatest concentration, C^{12} less, and $C^{13}N^{15}$ still less. (The concentration of the radioactive nuclei N^{13} and O^{15} is, of course, very small, about 10^{-12} of N^{14}). A comparison of the observed abundances of C and N is not very conclusive, because of the very different chemical properties. However, a comparison of the isotopes of each element should be significant.

In this respect, the result for the carbon isotopes is quite satisfactory. C^{13} captures slow protons about 70 times as easily as C^{12} (experimental value!), therefore C^{12} should be 70 times as abundant. The actual abundance ratio is 94 : 1. The same fact can be expressed in a more "experimental" way: In equilibrium, the number of reactions (42) per second should be the same as of (42b). Therefore, if a natural sample of terrestrial carbon (which is presumed to reproduce the solar equilibrium) is bombarded with protons, equally many captures should occur due to each carbon isotope. This is what is actually found experimentally;[18] the equality of the γ-ray intensities from C^{12} and C^{13} is, therefore, not accidental.[36]

The greater abundance of C^{12} is thus due to the smaller probability of proton capture which in turn appears to be due to the smaller $h\nu$ of

the capture γ-ray. Thus the great energetic stability of C^{12} actually makes this nucleus abundant. However, it is not because of a Boltzmann factor as has been believed in the past, but rather because of the small energy evolution of the proton capture reaction.

In nitrogen, the situation is different. Here N^{14} is energetically less stable (has higher mass excess) than N^{15} but is more abundant in spite of it (abundance ratio $\sim 500 : 1$). This must be due to the fact that N^{15} can give a $p-\alpha$ reaction while N^{14} can only capture a proton; particle reactions are always much more probable than radiative capture. Thus the greater abundance of N^{14} is due not to its own small mass excess but to the large mass excess of C^{11} which would be the product of the $p-\alpha$ reaction (43b).

Quantitative data on the nitrogen reactions (42c), (42e) are not available, the figures in Table V are merely estimates. If our theory about the abundance of the nitrogen isotopes is correct, the ratio of the reaction rates should be $500 : 1$, i.e., either $N^{14} + H = O^{15}$ must be more probable[36a] or $N^{15} + H = C^{12} + He^4$ less probable than assumed in Table V. Experimental investigations would be desirable.

Turning now to the energy evolution, we notice that the cycle (42) contains two radioactive processes (N^{13} and O^{15}) giving positrons with 1.3_5 and 1.8_5 mMU maximum energy, respectively. If we assume again that $\frac{2}{3}$ of the energy is, on the average, given to neutrinos, this makes 2.0 mMU neutrino energy per cycle, which is 7 percent of the total energy evolution (28.7 mMU). There are therefore $4.0 \cdot 10^{-5}$ ergs available from each cycle. (It may be mentioned that the neutrinos emitted have too low energy to cause the transformation of protons into neutrons according to (21).)

The duration of one cycle (42) is equal to the sum of the lifetimes of all nuclei concerned, i.e., practically to the lifetime of N^{14}. Thus each N^{14} nucleus will produce $4.0 \cdot 10^{-5}$ erg every $5 \cdot 10^7$ years, or $3 \cdot 10^{-20}$ erg per second. Under the

---

[36] It would be tempting to ascribe similar significance to the equality of intensity of the two α-groups from natural Li bombarded by protons ($Li^6 + H = He^4 + He^3$, $Li^7 + H = 2He^4$). However, the lithium isotopes do not seem to be genetically related as are those of carbon.

[36a] *Note added in proof:*—In this case, the life of N^{14} in the sun might actually be shorter, and its abundance smaller than that of C^{12}. Professor Russell pointed out to me that this would be in better agreement with the evidence from stellar spectra. Another consequence would be that a smaller abundance of N^{14} would be needed to explain the observed energy production.

TABLE VII. p–α reactions.

Initial Nucleus	Product Nucleus	Energy Evolution Q (MMU)	Potential Barrier B (MMU)	Penetrability P
Li^6	He^3	4.14		1
Li^7	He^4	18.59		1
Be^9	Li^6	2.45	2.6	1
B^{10}	Be^7	1.2	3.4$_5$	0.027
B^{11}	$2He^4$	9.4		1
C^{12}	B^9	−8.1		0
C^{13}	B^{10}	−4.4		0
N^{14}	C^{11}	−3.5		0
N^{15}	C^{12}	5.2	4.6	1
O^{16}	N^{13}	−5.8		0
O^{17}	N^{14}	1.3	5.1	$1.7 \cdot 10^{-4}$
O^{18}	N^{15}	3.1	5.0	0.2
F^{19}	O^{16}	8.8		1
Ne^{20}	F^{17}	−4.5		0
Ne^{21}	F^{18}	−1.6		0
Ne^{22}	F^{19}	−1.6		0
Na^{23}	Ne^{20}	1.6	6.7	$6 \cdot 10^{-6}$
Mg^{24}	Na^{21}	<−3.0		0
Mg^{25}	Na^{22}	−2.1		0
Mg^{26}	Na^{23}	−2.0		0
Al^{27}	Mg^{24}	1.8	7.6	$1.1 \cdot 10^{-6}$
Si^{28}	Al^{25}	<−2.9		0
Si^{29}	Al^{26}	−2.0		0
Si^{30}	Al^{27}	−2.4		0
P^{31}	Si^{28}	2.0	8.5	$2.5 \cdot 10^{-7}$
S^{32}	P^{29}	<0		0
S^{33}	P^{30}	−2		0
S^{34}	P^{31}	−2.0		0
Cl^{35}	S^{32}	2.3	9.3	$5 \cdot 10^{-8}$
Cl^{37}	S^{34}	4.2	9.1	$2.5 \cdot 10^{-3}$

assumption of a N^{14} concentration of 10 percent by weight, this gives an energy evolution of

$$\frac{6 \cdot 10^{23} \cdot 0.1}{14} \cdot 3 \cdot 10^{-20} \approx 100 \text{ ergs/g sec.} \quad (45)$$

at "standard stellar conditions," i.e., $T = 2 \cdot 10^7$, $\rho = 80$, hydrogen concentration 35 percent.

This result is just about what is necessary to explain the observed luminosity of the sun. Since the nitrogen reaction depends strongly on the temperature (as T^{18}) and the temperature, as well as the density, decrease rapidly from the center of the sun outwards, the average energy production will be only a fraction, perhaps 1/10 to 1/20, of the production at the center.[36b] This means that the average production is 5 to 10 ergs/g sec., in excellent agreement with the observed luminosity of 2 ergs/g sec.

[36b] *Added in proof:*—According to calculations of R. Marshak, the correct figure is about 1/30.

Thus we see that the reaction between nitrogen and protons which we have recognized as the logical reaction for energy production from the point of view of nuclear physics, also agrees perfectly with the observed energy production in the sun. This result can be viewed from another angle: We may ignore, for the moment, all our nuclear considerations and ask simply which nucleus will give us the right energy evolution in the sun? Or conversely: Given an energy evolution of 20 ergs/g sec. at the center of the sun, which nuclear reaction will give us the right central temperature ($\sim 19 \cdot 10^6$ degrees)?

This calculation has been carried out in Table VI. It has been assumed that the density is 80, the hydrogen concentration 35 percent and the concentration of the other reactant 10 percent by weight. The "widths" were assumed the same as in Table V. Given are the necessary temperatures for an energy production of 20 ergs/g sec. It is seen that all nuclei up to boron require extremely low temperatures in order not to give too much energy production; these temperatures ($< 10^7$ degrees) are quite irreconcilable with the equations of hydrostatic and radiation equilibrium. On the other hand, oxygen and neon would require much too high temperatures. Only carbon and nitrogen require nearly, and nitrogen in fact exactly, the central temperature obtained from the Eddington integrations ($19 \cdot 10^6$ degrees). Thus from stellar data alone we could have predicted that the capture of protons by N^{14} is the process responsible for the energy production.

§8. Reactions with Heavier Nuclei

Mainly for the sake of completeness, we shall discuss briefly the reactions of nuclei heavier than nitrogen. For the energy production, these reactions are obviously of no importance because the higher potential barrier of the heavier nuclei makes their reactions much less probable than those of the carbon-nitrogen group.

The most important point for a qualitative discussion is the question whether a $p - \alpha$ reaction is energetically possible for a particular nucleus, and, if possible, whether it is impeded by the potential barrier. In Table VII are listed the energy evolution in $p - \alpha$ reactions for all stable

세상에서 가장 쉬운 과학 수업 별의 물리학

(nonradioactive) nuclei up to chlorine. In the first column, the reacting nucleus is given, in the second, the product of a $p-\alpha$ reaction. The third column contains the reaction energy Q; when Q is negative, the reaction is energetically impossible so that the initial nucleus can only capture protons with γ-emission. In the fourth column, the height of the nuclear potential barrier is given for all reactions with positive Q. In the last column, the penetrability of the potential barrier is calculated according to standard methods (reference 10, p. 166). If $Q>B$, the penetrability is 1; if Q is negative, $P=0$ was inserted.

The *a priori* probability of a $p-\alpha$ reaction is roughly 10^4 times that of radiative capture. Therefore the emission of α-particles will be preferred when $P>10^{-4}$. It is seen from the table that for all nuclei up to boron the $p-\alpha$ reaction is strongly preferred, a fact which we recognized as the main reason for the impossibility of building up heavier elements than He^4 (§5). Furthermore, in the carbon group, only proton capture is possible for $C^{12}C^{13}N^{14}$ while for N^{15} the $p-\alpha$ reaction will strongly predominate (cf. §7).

The oxygen-fluorine group shows intermediate behavior. O^{16} can only capture protons, for O^{17} the capture and the α-emission will have roughly equal probability while for O^{18} and F^{19} the $p-\alpha$ reaction will be much more probable. With a ratio 10^4 for the *a priori* probabilities, about 40 percent of the O^{17} will become F^{18} (and then O^{18} by positron emission) while 60 percent will revert to N^{14}. Of the O^{18}, only 1 part in 2000 will become F^{19}, and of the F^{19}, only 1 in 10,000 will transform into Ne^{20}. Thus, under continued proton bombardment, about one O^{16} nucleus in $5 \cdot 10^7$ will ultimately transform into Ne^{20}, the rest will become nitrogen.

Actually, these considerations are somewhat academic because in general the supply of protons will be exhausted long before all the O^{16} initially present in the star will have captured a proton. Because of the small energy evolution in the reaction $O^{16}+H=F^{17}$, this reaction is extremely slow ($\sim10^{12}$ years) so that equilibrium in the oxygen group will not be reached in astronomical times.

Among the nuclei heavier than fluorine, the $p-\alpha$ reaction is in general energetically permitted only for those with mass number $4n+3$. But even for these, the energy evolution is so much less than the height of the barrier that the penetrabilities are extremely small. Thus for all these elements only proton capture will occur (with the possible exception of Cl^{37}).

These considerations demonstrate the uniqueness of the carbon-nitrogen cycle.

§9. Agreement with Observations

In Table VIII, we have made a comparison of our theory (carbon-nitrogen reaction) with observational data for five stars for which such data are given by Strömgren.[1] The first five columns are taken from his table, the last contains the necessary central temperature to give the correct energy evolution with the carbon-nitrogen reactions (cf. Table VI). As in §7, we have assumed a N^{14} content of 10 percent, and an energy production at the center of ten times the average energy production (listed in the second column).

The result is highly satisfactory: The temperatures necessary to give the correct energy evolution (last column) agree very closely with the temperatures obtained from the Eddington integration (second last column). The only exception from this agreement is the giant Capella: This is not surprising because this star has greater luminosity than the sun at smaller density and temperature; such a behavior cannot possibly be explained by the same mechanism which ac-

TABLE VIII. *Comparison of the carbon-nitrogen reaction with observations.*

STAR	LUMINOSITY ERG/G SEC.	CENTRAL DENSITY	H CONTENT (PER-CENT)	CENTRAL TEMPERATURE (MILLION DEGREES)	
				INTE-GRATION	ENERGY PRODUC-TION
Sun	2.0	76	35	19	18.5
Sirius A	30	41	35	26	22
Capella	50	0.16	35	6	32
U Ophiuchi (bright)	180	12	50	25	26
Y Cygni (bright)	1200	6.5	80	32	30

[37] A. S. Eddington, *The Internal Constitution of the Stars* (Cambridge University Press, 1926).

counts for the main sequence. We shall come back to the problem of energy production in giants at the end of this section.

For the main sequence we observe that the small increase of central temperature from the sun to Y Cygni (19 to $33 \cdot 10^6$ degrees) is sufficient to explain the much greater energy production (10^4 times) of the latter. The reason for this is, of course, the *strong temperature dependence of our reactions* ($\sim T^{18}$, cf. §10). We may say that *astrophysical data themselves would demand such a strong dependence* even if we did not know that the source of energy are nuclear reactions. The small deviations in Table VIII can, of course, easily be attributed to fluctuations in the nitrogen content, opacity, etc.

In judging the agreement obtained, it should be noted that the "observational" data in Table VIII were obtained by integration of an Eddington model,[1, 37] i.e., the energy production was assumed to be almost constant throughout the star. Since our processes are strongly temperature dependent, the "point source" model should be a much better approximation. However, it seems that the results of the two models are not very different so that the Eddington model may suffice until accurate integrations with the point source model are available.[37a]

Since our theory gives a definite mechanism of energy production, it permits decisions on questions which have been left unanswered by astrophysicists for lack of such a mechanism. The first is the question of the "model," which is answered in favor of one approximating a point source model. The second is the problem of chemical composition. The equilibrium conditions permit for the sun a hydrogen content of either 35 or 99.5 percent when there is no helium, and intermediate values when there is helium. The central temperature varies from $19 \cdot 10^6$ to $9.5 \cdot 10^6$ when the hydrogen content increases

from 35 to 99.5 percent. It is obvious that the latter value can be definitely excluded on the basis of our theory: The energy production due to the carbon-nitrogen reaction would be reduced by a factor of about 10^8 (100 for nitrogen concentration, 10^6 for temperature). The proton combination (1) would still supply about 5 percent of the observed luminosity; but apart from the fact that a factor 20 is missing, the proton combination does not depend sufficiently on temperature to explain the larger energy production in brighter stars of the main sequence. Thus it seems that only a small range of hydrogen concentrations around 35 percent is permitted; what this range is, depends to some extent on the N^{14} concentration and also requires a more accurate determination of the distribution of temperature and density.

Next, we want to point out a rather well-known difficulty about the energy production of very heavy stars such as Y Cygni. With an energy production of 1200 ergs/g sec., and an available energy of $1.0 \cdot 10^{-5}$ erg per proton (formation of α-particles!), *all* the energy will be consumed in $1.7 \cdot 10^8$ years. Since at present Y Cygni still has a hydrogen content of 80 percent, its past life should be less than $3.5 \cdot 10^7$ years. We must therefore conclude that Y Cygni and other heavy stars were "born" comparatively recently—by what process, we cannot say. This difficulty, however, is not peculiar to our theory of stellar energy production but is inherent in the well-founded assumption that nuclear reactions are responsible for the energy production.[38]

Finally, we want to come back to the problem of stars outside the main sequence. The white dwarfs presumably offer no great difficulty. The internal temperature of these stars is probably rather low,[1, 39] because of the low opacity (degeneracy!) so that the small energy production may be understandable. Quantitative calculations are, of course, necessary. For the giants, on the other hand, it seems to be rather difficult to account for the large energy production by nuclear reactions. If the Eddington (or the point

---

[37a] Mr. Marshak has kindly calculated the central temperature and density of the sun for the point source model, using Strömgren's tables for which we are indebted to Professor Strömgren. With an average atomic weight $\mu = 1$, Marshak finds

for the point source model $T_c = 20.3 \cdot 10^6$, $\rho_c = 50.2$
for the Eddington model $T_c = 19.6 \cdot 10^6$, $\rho_c = 72.2$

Not only is the temperature difference very small ($3\frac{1}{2}$ percent) but it is, for the sake of the energy production, almost compensated by a density difference in the opposite direction. The product $\rho_c T_c^{18}$ is only 20 percent greater for the point source model.

[38] Even if the most stable nuclei (Fe, etc.) are formed rather than He, the possible life will only increase by 30 percent.

[39] S. Chandrasekhar, Monthly Not. **95**, 207, 226, 676 (1935).

세상에서 가장 쉬운 과학 수업 별의 물리학

TABLE IX. *Relations between stellar constants.*

Quantity	$\gamma=18$ $\alpha=\frac{1}{2}$ $\beta=2\frac{3}{4}$ $\delta=19\frac{3}{4}$	$\gamma=18$ $\alpha=1$ $\beta=3\frac{1}{2}$ $\delta=20\frac{1}{2}$	$\gamma=15$ $\alpha=\frac{1}{2}$ $\beta=2\frac{3}{4}$ $\delta=16\frac{3}{4}$	$\gamma=4.5$ $\alpha=\frac{1}{2}$ $\beta=2\frac{3}{4}$ $\delta=6\frac{1}{4}$
Radius R	$M^{0.75}\mu^{0.57}(yz)^{0.05}$	$M^{0.71}\mu^{0.51}(yz)^{0.05}$	$M^{0.70}\mu^{0.49}(yz)^{0.06}$	$M^{0.20}\mu^{-0.36}(yz)^{0.16}$
Temperature T	$M^{0.25}\mu^{0.43}(yz)^{-0.05}$	$M^{0.29}\mu^{0.49}(yz)^{-0.06}$	$M^{0.30}\mu^{0.51}(yz)^{-0.06}$	$M^{0.80}\mu^{1.36}(yz)^{-0.16}$
Density ρ	$M^{-1.25}\mu^{-1.71}(yz)^{-0.15}$	$M^{-1.13}\mu^{-1.53}(yz)^{-0.15}$	$M^{-1.10}\mu^{-1.47}(yz)^{-0.18}$	$M^{0.40}\mu^{1.08}(yz)^{-0.48}$
Luminosity L	$M^{4.32}\mu^{6.04}y^{-1.06}z^{-0.06}$	$M^{5.15}\mu^{7.24}y^{-1.02}z^{-0.02}$	$M^{4.37}\mu^{6.13}y^{-1.07}z^{-0.07}$	$M^{6.00}\mu^{7.20}y^{-1.20}z^{-0.20}$
Surf. temp. T_S	$M^{0.70}\mu^{1.23}y^{-0.29}z^{-0.04}$	$M^{0.93}\mu^{1.85}y^{-0.28}z^{-0.03}$	$M^{0.74}\mu^{1.29}y^{-0.30}z^{-0.05}$	$M^{1.25}\mu^{1.98}y^{-0.28}z^{-0.13}$

source) model is used, the central temperatures and densities are exceedingly low, e.g., for Capella $T=6\cdot10^6$, $\rho=0.16$. Only a nuclear reaction going at very low temperature is therefore at all possible; $Li^7+H=2He^4$ would be just sufficient. But it seems difficult to conceive how the Li^7 should have originated in *all* the giants in the first place, and why it was not burned up long ago. The only other source of energy known is gravitation, which would require a core model[40] for giants.[41] However, any core model seems to give small rather than large stellar radii.

§10. The Mass-Luminosity Relation

In this section, we shall use our theory of energy production to derive the relation between mass and luminosity of a star. For this purpose, we shall employ the well-known homology relations (reference 1, p. 492). This is justified because we assume that all stars have the same mechanism of energy evolution and therefore follow the same model. Further, it is assumed that the matter throughout the star is non-degenerate which seems to be true for all stars except the white dwarfs. (For all considerations in this and the following two sections, cf. reference 1.)

We shall consider the mass of the star M, the mean molecular weight μ, the concentration of "Russell Mixture" y and the product of the concentrations of hydrogen and nitrogen, z, as independent variables. In addition, we introduce for the moment the radius R which, however, will be eliminated later. Then, obviously, we have for the density (at each point)

$$\rho\sim M/R^3. \qquad (46)$$

From the equation of hydrostatic equilibrium

$$dp/dr=-GM_r\rho/r^2 \qquad (47)$$

(G=gravitational constant) and the gas equation

$$p=RT\rho/\mu \qquad (47a)$$

(R the gas constant, radiation pressure neglected), we find

$$T\sim M\mu/R. \qquad (48)$$

Finally, we must use the equation of radiative equilibrium:

$$aT^3\frac{dT}{dr}=-\frac{3}{4}\frac{k}{c}\rho\frac{L_r}{4\pi r^2}, \qquad (49)$$

where a is the Stefan-Boltzmann constant, c the velocity of light, k the opacity, and L_r the luminosity (energy flux) at distance r from the center.

For the opacity, we assume

$$k\sim y\rho^\alpha T^{-\beta} \qquad (50)$$

(y concentration of heavy elements). Usually, α is taken as 1 and $\beta=3.5$ (Kramers' formula). However, the Kramers formula must be divided by the "guillotine factor" τ which was calculated from quantum mechanics by Strömgren.[42] For densities between 10 and 100, and temperatures between 10^7 and $3\cdot10^7$, Strömgren's numerical results can be fairly well represented by taking $\tau\sim\rho^{\frac{1}{3}}T^{-\frac{1}{3}}$. Therefore we adopt $\alpha=\frac{1}{2}$, $\beta=2.75$ in (50).

The luminosity may be written

$$L\sim M\rho z T^\gamma. \qquad (51)$$

That the energy production per unit mass, L/M, contains a factor ρ follows from the fact that it is due to two-body nuclear reactions; this factor is

---

[40] L. Landau, Nature **141**, 333 (1938).
[41] This suggestion was made by Gamow in a letter to the author.

[42] Cf. reference 1, Table 6, p. 485.

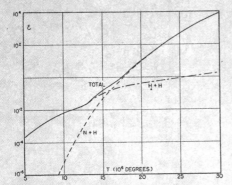

FIG. 1. The energy production in ergs/g sec. due to the proton-proton combination (curve H+H) and the carbon-nitrogen cycle (N+H), as a function of the central temperature of the star. Solid curve: total energy production caused by both reactions. The following assumptions were made: central density = 100, hydrogen concentration .35 percent, nitrogen 10 percent; average energy production 1/5 of central production for H+H, 1/10 for N+H.

apparent from all our formulae, e.g., (4). $z = x_1x_2$ is the product of the concentrations of the reacting nuclei (N^{14} and H). For γ we obtain from (4), (6)

$$\gamma = \frac{d \log (\tau^2 e^{-\tau})}{d \log T} = \tfrac{1}{3}(\tau - 2).$$ (52)

For $N^{14}+H$ and $T = 2 \cdot 10^7$, this gives $\gamma = 18$. For $T = 3.2 \cdot 10^7$ (Y Cygni), $\gamma = 15.5$; generally, $\gamma \sim T^{-\frac{1}{3}}$. $\gamma = 18$ will be a good approximation over most of the main sequence.

Inserting (50), (51) in (49), we have

$$T^{4+\beta-\gamma} \sim yzM\rho^{2+\alpha}R^{-1}.$$ (53)

Combining this with (46), (48), and introducing the abbreviation

$$\delta = \gamma + 3 + \tfrac{3}{2}\alpha - \beta$$ (54)

we find

$$R \sim M^{1-2(2+\alpha)/\delta}\mu^{1-(7+\delta\alpha)/\delta}(yz)^{1/\delta},$$ (55)

$$T \sim M^{2(2+\alpha)/\delta}\mu^{(7+3\alpha)/\delta}(yz)^{-1/\delta},$$ (56)

$$\rho \sim M^{-2+6(2+\alpha)/\delta}\mu^{-3+3(7+3\alpha)/\delta}(yz)^{-3/\delta},$$ (57)

$$L \sim M^{3+2\alpha+2(2+\alpha)(\beta-3\alpha)/\delta}\mu^{4+3\alpha+(7+3\alpha)(\beta-3\alpha)/\delta}$$
$$\times y^{-1}(yz)^{-(\beta-3\alpha)/\delta}.$$ (58)

Furthermore, the surface temperature may be of interest, we have

$$T_S \sim L^{\frac{1}{4}}R^{-\frac{1}{2}} \sim M^{\frac{1}{4}+\frac{1}{2}\alpha+\frac{1}{2}(2+\alpha)(\beta-3\alpha+2)/\delta}$$
$$\times \mu^{\frac{1}{4}+\frac{3}{4}\alpha+\frac{1}{4}(7+3\alpha)(\beta-3\alpha+2)/\delta}y^{-\frac{1}{4}}$$
$$\times (yz)^{-\frac{1}{4}(\beta-3\alpha+2)/\delta}.$$ (59)

In Table IX, these formulas are given explicitly, for four different sets of constants.

The most important result is that the central *temperature* depends only slightly on the mass of the star, *viz.* as $M^{0.25}$ and $M^{0.30}$ for $\gamma = 18$ and 15. The reason for this is the strong temperature dependence of the reaction rate: The exponent of M in (56) is inversely proportional to δ which is mainly determined (cf. (54)) by the exponent γ in formula (51) for the temperature dependence of the reaction rate. The integration of the Eddington equations with the use of observed luminosities, radii, etc., gives, in fact, only a small dependence of the central temperature on the mass. This can *only* be explained by a strong temperature dependence of the source of stellar energy, a fact which has not been sufficiently realized in the past. Theoretically, the central temperature increases somewhat with increasing mass of the star, more strongly with the mean molecular weight, and is practically independent of the chemical composition, i.e., of y and z.

The *radius* of the star is larger for heavy stars and for high molecular weight. The density behaves, of course, in the opposite way. Both these results are in qualitative agreement with observation. The product of mass and density which occurs in Eq. (51) for the luminosity, is almost independent of the mass; therefore, for constant concentrations z, the luminosity is determined by the central temperature alone. Both radius and density are almost independent of the chemical composition, except insofar as it affects μ.

The *luminosity* increases slightly faster than the fourth power of the mass and the sixth power of the mean molecular weight. This increase is considerably less than that usually given ($M^{5.5}\mu^{7.5}$) and agrees better with observation. The difference from the usual formula is mainly due to the different dependence of the opacity on density and temperature; in fact, with the usual assumption ($\alpha = 1$, $\beta = 3\tfrac{1}{2}$), we obtain $M^{5.15}\mu^{7.24}$.

세상에서 가장 쉬운 과학 수업 별의 물리학

The remaining difference is that the dependence on the radius is carried as a separate factor in the usual formula while we have expressed R in terms of M and μ. The observations show, for bright stars, an even slower increase than M^4; this seems to be due to the lower average molecular weight (higher hydrogen content) of most very bright stars. It may be that these stars become unstable because of excessive energy production when their hydrogen content becomes too low (cf. §12).—The luminosity is inversely proportional to the concentration y of heavy elements (Russell mixture) because y determines the opacity. However, L is almost independent of the nitrogen concentration, as are R, T and ρ.

All these considerations are valid for bright stars, down to about three magnitudes fainter than the sun. For fainter stars, with lower central temperatures, the proton combination $H+H = D+\epsilon^+$ should become more probable than the carbon-nitrogen reactions, because this reaction depends less on temperature. In discussing the energy production from $H+H$, we must take into account that (cf. §5) at low temperatures, this reaction leads only to He^3 rather than He^4, because of the slowness of the reaction $He^4+He^3 = Be^7$. (Assumption B of §5 is made, viz. that He^3 is more stable than H^3, and Li^4 unstable.) This causes a rather sudden drop in the energy evolution from $H+H$ around $14 \cdot 10^6$ degrees (cf. Fig. 1), i.e., just below the temperature at which the proton combination becomes important ($\sim 16 \cdot 10^6$ degrees, see Fig. 1). Therefore, the temperature exponent γ stays fairly large (~ 13, cf. Fig. 2) down to about $13 \cdot 10^6$ degrees which corresponds to an energy production of about one percent of that of the sun (five magnitudes fainter). For still fainter stars, γ drops to very low values, reaching a minimum of about 4.5 near 10^7 degrees.

The relations between central temperature, radius, luminosity and mass for this case ($\gamma = 4.5$) are given in the last column of Table IX. The temperature is seen to depend much more strongly on the mass (as $M^{0.80}$) while the radius becomes almost independent of M and the density decreases with decreasing mass. The luminosity decreases faster with the mass (as M^5) than for the bright stars. Unfortunately, little material is available for these fainter stars. This is the more regrettable as it is rather important for nuclear physics to decide whether the $H+H$ reaction is really as probable as assumed in this paper: There is some possibility that it is forbidden by selection rules (cf. reference 16, p. 250) in which case it would be about 10^6 times less probable. Then the carbon-nitrogen reactions would furnish the energy even for faint stars, and the central temperature of these stars would not depend much on their mass.

Figure 1 gives the energy production due to the proton combination ($H+H$) and to the carbon-nitrogen reactions ($N+H$) as a function of the central temperature, on a logarithmic scale. The great preponderance of $H+H$ at low and $N+H$ at high temperatures is evident. The following assumptions were made: Hydrogen concentration 35 percent, nitrogen 10 percent, central density $\rho = 100$, average energy production $= \frac{1}{3}$ of central production for $H+H$, $\frac{1}{10}$ for $N+H$ reaction. He^3 is supposed to be more stable than H^3, and Li^4 unstable (assumption B of §5).

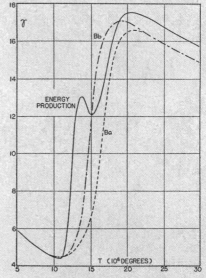

Fig. 2. The exponent γ in the relation $L \sim T^\gamma$ between luminosity L and temperature T, as a function of T. Solid curve: γ for total energy production (logarithmic derivative of solid curve in Fig. 1). Dotted curves: γ for stability against temperature changes (curve Ba for times less than 14 months, Bb for more than 14 months).

Figure 2, solid curve, gives the exponent γ in the *total* energy production. It is low at low temperatures ($\sim$4–5, hydrogen reaction) and has a minimum of 4.44 at 11 million degrees. Between 11 and 14 million degrees, γ rises steeply as the reaction $He^4 + He^3 = Be^7$ sets in (§5); then it falls again from 13 to 12 when this reaction reaches saturation. From $T = 15$, the carbon-nitrogen reactions set in, causing a rise of γ to a maximum of 17.5 at $T = 20$, while at higher temperatures γ decreases again as $T^{-\frac{1}{3}}$.

§11. Stability Against Temperature Changes

Cowling[43] has investigated the stability of stars against vibration. This stability is determined mostly by the ratio γ of the specific heats at constant pressure and constant volume. If the radiation pressure in the star is negligible compared with the gas pressure ($\gamma = 5/3$) then the star will be stable for any value of the temperature exponent n in the energy production (38), up to $n \approx 450$. Only for very heavy stars, for which the radiation pressure is comparable with the gas pressure, does the stability condition put a serious restriction on the temperature dependence of the energy production. According to our theory, the energy production is proportional to T^{17} (see below); according to Cowling, stability will then occur when $\gamma > 10/7$. The corresponding ratio of radiation pressure to total pressure is

$$1 - \beta = \frac{5 - 3\gamma}{3(7\gamma - 9)} = 0.24 \qquad (60)$$

and the corresponding mass of the star, according to Eddington's "standard model," is $17/\mu^2\,\odot$, where μ is the average molecular weight. Therefore practically all the stars for which good observational data are available will be stable.

It may be worth while to point out that the temperature exponent n for these stability considerations is not exactly equal to that for the energy production in equilibrium. A change of temperature gives rise to radial vibrations of the star whose period is of the order of days or, at most, a few years. On the other hand, when the temperature is raised or lowered, the equilibrium

[43] T. G. Cowling, Monthly Not. **94**, 768 (1934); **96**, 42 (1935).

between carbon and nitrogen will be disturbed and it takes a time of the order of the lifetime of C^{12} ($\sim 10^6$ years) to restore equilibrium at the new temperature. Thus we must take the concentrations of carbon and nitrogen corresponding to the *original* temperature.

At $T = 2 \cdot 10^7$ degrees, the carbon reactions have a $\tau = 50.6$ (Table V) and therefore a temperature exponent $\gamma_C = 16.2$ (cf. 52); the nitrogen reactions have $\tau = 56.3$ and $\gamma_N = 18.1$. The number of reactions per second is, in equilibrium, the same for each of the reactions in the chain (42). The energy evolution in the first three reactions together is 11.7 mMU, after subtracting 0.8_5 mMU for the neutrino emitted by N^{13}. These three reactions are carbon reactions, the remaining three, with an energy evolution of 15.0 mMU, are nitrogen reactions. Thus the effective temperature exponent for stability problems becomes

$$\gamma = \frac{11.7\gamma_C + 15.0\gamma_N}{26.7} = 17.3 \qquad (60a)$$

at $2 \cdot 10^7$ degrees. γ is approximately proportional to $T^{-\frac{1}{3}}$ (cf. 52, 6).

For the proton combination, we have to distinguish the three possibilities discussed in §5. The simplest of these is assumption C.

Assumption C: Li^4 unstable, H^3 more stable than He^3

In this case, there are two "slow" processes, *viz.* (a) the original reaction $H + H$ and (b) the transformation of He^3 into H^3 by electron capture ($\sim$2000 years). (a) depends on temperature approximately as $T^{3.5}$, (b) is independent of T. The energy evolution up to the formation of He^3 is 7.2 mMU, from He^3 to He^4 21.3 mMU. Thus

$$\gamma = \frac{7.2 \cdot 3.5 + 21.3 \cdot 0}{28.5} = 0.9. \qquad (61C)$$

This would be a very slight dependence indeed.

Assumption A: Li^4 stable

According to Table V, the transformation of He^3 into Li^4 takes about one day.

(a) For times *shorter than one day*, the reactions up to He^3 and from then on are independent of each other. The first group again has $\gamma_H = 3.5$ and gives 7.2 mMU, the second group now gives only 8.3 mMU because the remaining 13.0 are lost to the neutrino from Li^4 (cf. §5) and has (Table V) $\gamma_{He} = 6.9$. Therefore

$$\gamma = \frac{7.2 \cdot 3.5 + 8.3 \cdot 6.9}{15.5} = 5.3. \qquad (61Aa)$$

세상에서 가장 쉬운 과학 수업 별의 물리학

(b) For times *longer than one day*, the proton combination determines the whole chain of reactions so that

$$\gamma = 3.5. \qquad (61Ab)$$

Assumption B. Li⁴ unstable, He³ more stable than H³ (most probable assumption)

As was pointed out in §5, the reaction $He^4 + He^3 = Be^7$ is so slow ($3 \cdot 10^7$ years, Table V) that the concentration of He^3 will remain unaffected by the temperature fluctuation. Be^7 has a mean life of 14 months at the center of the sun so that there are again two cases:

(a) *For times less than 14 months*, there are three groups of reactions, (a) those up to He^3, giving again 7.2 mMU with $\gamma_H = 3.5$, (b) the reaction $He^4 + He^3 = Be^7$, giving (Table V) 1.6 mMU with $\gamma_{He} = 15.1$, and (c) the electron capture by Be^7, followed by $Li^7 + H = 2He^4$. This last reaction does not depend on temperature ($\gamma_{Be} = 0$) and gives 18.6 mMU. Therefore the effective γ becomes

$$\gamma = \frac{7.2 \cdot 3.5 + 1.6 \cdot 15.1 + 18.6 \cdot 0}{27.4} = 1.8. \qquad (61Ba)$$

(b) *For times longer than 14 months*, the reaction $He^4 + He^3 = Be^7$ governs all the energy evolution from He^3 on, so that

$$\gamma = \frac{7.2 \cdot 3.5 + 20.2 \cdot 15.1}{27.4} = 12.1 \qquad (61Bb)$$

at $2 \cdot 10^7$ degrees.

At low temperatures, the reaction $He^4 + He^3 = Be^7$ stops altogether (cf. §5) so that then the γ of the H+H reaction itself determines the radial stability. From 12 to $16 \cdot 10^6$ degrees we have a transition region in which the importance of the $He^4 + He^3$ reaction (and the consequent ones) is reduced.

In Fig. 2, curves Ba and Bb, we have plotted the effective γ for the radial stability, by taking into account both C+N and H+H reactions, and making the same assumptions about the concentrations of hydrogen and nitrogen as in Fig. 1 (cf. end of §10). Assumption B was made regarding the stability of Li⁴ and He³; the curves Ba and Bb refer to the short time and long time formulas (61Ba) and (61Bb). At high temperatures, the two curves coincide because

then the proton combination is unimportant compared with the carbon-nitrogen reactions. The combined γ is seen to reach a maximum of 17 (for the long time curve).

§12. STELLAR EVOLUTION[44]

We have shown that the concentrations of heavy nuclei (Russell mixture) and, therefore, also of nitrogen, cannot change appreciably during the life of a star. The only process that occurs is the transformation of hydrogen into helium, regardless of the detailed mechanism. The state of a star is thus described by the hydrogen concentration x, and by a *fixed* parameter, y, giving the concentration of Russell mixture. The rest, $1 - x - y$, is the helium concentration. Without loss of generality, we may fix the zero of time so that the helium concentration is zero. (Then the actual "birth" of the star may occur at $t > 0$).

It has been shown (Table IX) that the luminosity depends on the chemical composition practically only[45] through the mean molecular weight μ. This quantity is given by

$$1/\mu = 2x + \tfrac{3}{4}(1 - x - y) + \tfrac{1}{2}y = (5/4)(x + a), \quad (62)$$

$$a = 0.6 - 0.2y, \qquad (62a)$$

taking for the molecular weights of hydrogen, helium and Russell mixture the values $\tfrac{1}{2}$, 4/3 and 2, respectively.[1]

Now the rate of decrease of the hydrogen concentration is proportional to the luminosity, which we put proportional to μ^n. According to Table IX, n is about 6. Then

$$dx/dt \sim -(x + a)^{-n}. \qquad (63)$$

Integration gives

$$(x + a)^{n+1} = A(t_0 - t), \qquad (64)$$

where A is a constant depending on the mass and other characteristics of the star. Since $x = 1 - y$ at $t = 0$, we have

$$At_0 = (1.6 - 1.2y)^{n+1}. \qquad (64a)$$

---

[44] Most of these considerations have already been given by G. Gamow, Phys. Rev. **54**, 480(L) (1938).
[45] Except for the factor y^{-1} which, however, does not change with time.

It is obvious from (63) and (64) that the hydrogen concentration decreases slowly at first, then more and more rapidly. E.g., when the concentration of heavy elements is $y=\frac{1}{2}$, the first half of the hydrogen in the star will be consumed in 87 percent of its life, the second half in the remaining 13 percent. If the concentration of Russell mixture is small, the result will be even more extreme : For $y=0$, it takes 92 percent of the life of the star to burn up the first half of the hydrogen. Consequently, very few stars will actually be found near the end of their lives even if the age of the stars is comparable with their total lifespan t_0 (cf. 64a). In reality, the lifespan of all stars, except the most brilliant ones, is long compared with the age of the universe as deduced from the red-shift ($\sim 2 \cdot 10^9$ years) : E.g., for the sun, only one percent of the total mass transforms from hydrogen into helium every 10^9 years so that there would be only 2 percent He in the sun now, provided there was none "in the beginning." The prospective future life of the sun should according to this be $12 \cdot 10^9$ years.

It seems to us that this comparative youth of the stars is one important reason for the existence of a *mass*-luminosity relation—if the chemical composition, and especially the hydrogen content, could vary absolutely at random we should find a greater variability of the luminosity for a *given* mass.

It is very interesting to ask what will happen to a star when its hydrogen is almost exhausted. Then, obviously, the energy production can no longer keep pace with the requirements of equilibrium so that the star will begin to contract.

(This is, in fact, indicated by the factor $z^{1/6}$ in Eq. (55) for the stellar radius ; z is proportional to the hydrogen concentration.) Gravitational attraction will then supply a large part of the energy. The contraction will continue until a new equilibrium is reached. For "light" stars of mass less than $6\mu^{-2}$ sun masses (reference 1, p. 507), the electron gas in the star will become degenerate and a white dwarf will result. In the white dwarf state, the necessary energy production is extremely small so that such a star will have an almost unlimited life. This evolution was already suggested by Strömgren.[1]

For heavy stars, it seems that the contraction can only stop when a neutron core is formed. The difficulties encountered with such a core[46] may not be insuperable in our case because most of the hydrogen has already been transformed into heavier and more stable elements so that the energy evolution at the surface of the core will be by gravitation rather than by nuclear reactions. However, these questions obviously require much further investigation.

These investigations originated at the Fourth Washington Conference on Theoretical Physics, held in March, 1938 by the George Washington University and the Department of Terrestrial Magnetism. The author is indebted to Professors Strömgren and Chandrasekhar for information on the astrophysical data and literature, to Professors Teller and Gamow for discussions, and to Professor Konopinski for a critical revision of the manuscript.

[46] G. Gamow and E. Teller, Phys. Rev. **53**, 929(A), 608(L) (1938).

논문 웹페이지

XLVIII. *The Density of White Dwarf Stars.*

By S. CHANDRASEKHAR *.

Feb. 1931

1. THE first application of the Fermi-Dirac statistics to stellar problems was by Fowler † in connexion with the well-known problem of the companion of Sirius. This idea has lately been taken up by Stoner ‡ and others to calculate the limiting density of white dwarf stars. In this paper another way of arriving at the order of magnitude of the density of white dwarfs from different considerations is given.

2. Let p_r denote the radiation pressure and p_G the gas pressure, and the total pressure P is then given by

$$P = p_r + p_G. \quad \cdot \quad \cdot \quad \cdot \quad \cdot \quad \cdot \quad (1)$$

We introduce the constant β, such that

$$\left.\begin{array}{l} p_r = (1-\beta)P, \\ p_G = \beta P. \end{array}\right\} \quad \cdot \quad \cdot \quad \cdot \quad \cdot \quad (2)$$

We will make the assumption that $\beta = 1$ approximately, *i.e.,* we leave the radiation pressure out of account. We are dealing therefore with *ideal* conditions which can perhaps exist only in stars which are much higher in the white dwarf stage than even O_2, Eridani B.

---

* Communicated by R. H. Fowler, F.R.S.

† R. H. Fowler. Month. Not. Roy. A. S. lxxxvii. p. 114 (1926)

‡ E. C. Stoner, Phil. Mag. vii. p. 63 (1929): ix. p. 944 (1930).

Now for a fully degenerate electron gas (in the Sommerfeld sense) the pressure is given by

$$p_e = \frac{\pi}{60} \frac{h^2}{m} \left(\frac{3n}{\pi}\right)^{5/3} \quad \cdots \quad \cdots \quad (3)$$

We assume that it is this electron pressure which is by far the greatest contribution to the gas pressure, und therefore to the total pressure. Further, if ρ is the density of the stellar material, the number of electrons is given by

$$n = \frac{\rho}{\mu H(1+f)}, \quad \cdots \quad \cdots \quad (4)$$

where f is the ratio of the number of ions to the number of electrons (we can in practice neglect f), H the mass of the hydrogen atom, and μ the molecular weight. For a fully ionized material of the type we are dealing with $\mu = 2 \cdot 5$ nearly. We will use this value Later. We have therefore

$$p_G = \frac{\pi}{60} \frac{h^2}{m} \left(\frac{3}{\pi H}\right)^{5/3} \frac{\rho^{5/3}}{\mu^{5/3}(1+f)^{5/3}}$$

$$= 9 \cdot 845 \times 10^{12} \left[\frac{\rho}{\mu(1+f)}\right]^{5/3}, \quad \cdots \quad (5)$$

(The values used for h, m, etc. are those given in. A. S. Eddington's 'Internal Constitution of Stars,' Appendix (1).)
Putting

$$K = \frac{9 \cdot 845 \times 10^{12}}{\mu^{5/3}(1+f)^{5/3}}, \quad \cdots \quad \cdots \quad (6)$$

we have for the total pressure

$$P = K\rho^{5/3}. \quad \cdots \quad \cdots \quad (7)$$

We can now straightway apply the theory of the polytropic gas spheres, where for the exponent γ we have

$$\gamma = 5/3 \text{ or } 1 + \frac{1}{n} = 5/3,$$

giving $\qquad n = 3/2. \quad \cdots \quad \cdots \quad \cdots \quad (8)$
We have therefore*

$$\left(\frac{GM}{M'}\right)^{+1/2} \left(\frac{R'}{R}\right)^{-3/2} = \frac{[5/2K]^{3/2}}{4\pi G}, \quad \cdots \quad (9)$$

* A. S. Eddington, 'Internal Constitution of Stars,' p. 83 *et seq*. The notation is the same as that used in his book and now generally adopted. The particular equation (9) follows from the second of the equations (57.3)

세상에서 가장 쉬운 과학 수업 별의 물리학

Or $\dfrac{GM}{M'} = \dfrac{125 \times 9{\cdot}845^3 \times 10^{36}}{128\pi^2 G^3} \cdot \dfrac{1}{\mu^9 (1+f)^6} \left(\dfrac{R'}{R}\right)^3$ . . (10)

The values of R' and M' can be obtained from the extensive tables given by Emden in his 'Gas-Kugeln,' and are (page 79, tabbelle 4)

$$\left.\begin{aligned} R' &= 3{\cdot}6571, \\ M' &= 2{\cdot}7176. \end{aligned}\right\} \cdot \quad \cdot \quad \cdot \quad \cdot \quad \cdot \quad \cdot \quad (11)$$

Using these values in (10), and expressing the mass in terms of that of the Sun ($= 1{\cdot}985 \times 10^{33}$ grams), we get the result

$$(M/\odot)R^3 = \dfrac{2{\cdot}14 \times 10^{28}}{\mu^5} (= 2{\cdot}192 \times 10^{26}). \quad . \quad (12)$$

The second value for $(M/\odot)$ R^3, given in brackets, we get by using the value $2{\cdot}5$ for μ. We can express (12) differently, as follows:

$$R^6\rho = \dfrac{1{\cdot}014 \times 10^{61}}{\mu^5} (= 1{\cdot}039 \times 10^{59}), \quad . \quad . \quad (13)$$

$$\rho = 2{\cdot}162 \times 10^6 (M/\odot)^2. \quad . \quad . \quad . \quad (14)$$

We will apply the above equations to the case of the companion of Sirius. The mass of it, as determined from the double star orbit, is trustworthy, and equals $\cdot 85\odot$. The computed radius $= 1{\cdot}8 \times 10^7$. (But we cannot use this value in (13) to calculate the density, as it is based on formulae which may not be applicable to this case.) From the mass we can derive the radius und equals $6{\cdot}361 \times 10^8$ (about thirty times the accepted value). For the density of the companion of Sirius we get from (14), *provided* it were completely degenerate (which, however, is extremely unlikely),

$$\rho c._{\text{ Sirius}} = 1{\cdot}562 \times 10^6 \text{ grams per cm.}^3 \quad . \quad (15)$$

The mean density assumed is $\cdot 5 \times 10^5$, being thus thirty times smaller than that given by (15). We can, however, take the value given by (15) as indicating the *maximum* density which a stellar material having a mass equal to that of the companion of Sirius can have. A similar calculation can be made for O_2 Eridani B and Procyon B, and the calculated values are collected in a table below. The calculations for the *limiting density* on Stoner's theory give different values, and they are also given for comparison. We discuss the cause of the difference below.

We further note (i.) that the radius of a white dwarf is inversely proportional to the cube root of the radius, (ii.) the density is proportional to the square of the mass, (iii.) the central density would be six times the mean density ρ.

3. Stoner (*loc. cit.*) arrives at a formula for the *limiting* density for a material composed of completely ionized atoms on the following argument:—

The density increases as the sphere shrinks, and the limit is reached when the gravitational energy released just supplies the "energy required to squeeze the electrons closer together." The limiting condition would then be given by

$$\frac{d}{dn}(E_G + E_K) = 0, \quad \cdots \cdots \quad (16)$$

E_G being the gravitational energy and E_K the kinetic energy, for which, of course, the Fermi formula is used. The formula he gets is (without his latter relativity-mass correction)

$$\rho_{max.} = 3 \cdot 977 \times 10^6 \, (M/\odot)^2, \quad \cdots \quad (17)$$

which is exactly the same as our (14) with a difference in the

Star.	Mass.	Radius.	Density.		
			As calc. by (14).	Accepted value.	By Stoner's formula (17).
O$_2$ Eridani B.	·44 $\odot$	$7 \cdot 927 \times 10^8$	$4 \cdot 186 \times 10^5$	·98 $\times 10^5$	$7 \cdot 8 \times 10^5$
Procyon B.	·37 $\odot$	$8 \cdot 399 \times 10^8$	$2 \cdot 960 \times 10^5$	—	$5 \cdot 445 \times 10^5$
Companion of Sirius. }	·85 $\odot$	$6 \cdot 361 \times 10^8$	$1 \cdot 562 \times 10^6$	·5 $\times 10^5$	$2 \cdot 872 \times 10^6$

numerical factor only, the discrepancy being about 1:2. The difference in the two is obviously due to the fact that our value for ρ is not the " limiting density" in the sense in which Stoner uses the term ; but our calculation gives us a much nearer approximation to the conditions actually existent in white dwarfs than Stoner's calculation does. At any rate, it brings out clearly that the *order of magnitude* of the density which one can on purely theoretical considerations attribute to a white dwarf is the same.

Our results (see table) agree with Stoner's in showing that O$_2$ Eridani B is much nearer the ideal dwarf-star stage than the companion of Sirius, but indicate also that neither of them is so far from the ideal stage as Stoner's calculation would seem to indicate.

세상에서 가장 쉬운 과학 수업 별의 물리학

Summary.

The density of the white dwarf stars is reconsidered from the point of view of the theory of poly tropic gas spheres, and gives for the *mean density* of a white dwarf (under ideal conditions) the formula

$$\rho = 2 \cdot 162 \times 10^6 \times (M/\odot)^2.$$

The above formula is derived on considerations which are a much nearer approximation to the conditions *actually existent* in a white dwarf than the previous calculations of Stoner based on uniform density distribution in the star and which gave for the limiting density the formula

$$\rho = 3 \cdot 977 \times 10^6 \times (M/\odot)^2.$$

논문 웹페이지

THE MAXIMUM MASS OF IDEAL WHITE DWARFS

By S. CHANDRASEKHAR

ABSTRACT

The theory of the *polytropic gas spheres* in conjunction with the equation of state of a *relativistically degenerate electron-gas* leads to a *unique value for the mass of a star* built on this model. This mass (= $_{0.91}\odot$) is interpreted as representing the upper limit to the mass of an ideal white dwarf.

In a paper appearing in the *Philosophical Magazine*,[1] the author has considered the density of white dwarfs from the point of view of the theory of the polytropic gas spheres, in conjunction with the degenerate non-relativistic form of the Fermi-Dirac statistics. The expression obtained for the density was

$$\rho = 2.162 \times 10^6 \times \left(\frac{M}{\odot}\right)^2 , \tag{I}$$

where $M/\odot$ equals the mass of the star in units of the sun. This formula was found to give a much better agreement with facts than the theory of E. C. Stoner,[2] based also on Fermi-Dirac statistics but on uniform distribution of density in the star which is not quite justifiable.

In this note it is proposed to inquire as to what we are able to get when we use the relativistic form of the Fermi-Dirac statistics for the degenerate case (an approximation applicable if the number of electrons per cubic centimeter is $> 6 \times 10^{29}$). The pressure of such a gas is given by (which can be shown to be rigorously true)

$$P = \tfrac{1}{8}\left(\frac{3}{\pi}\right)^{\tfrac{1}{3}} \cdot hc \cdot n^{4/3} , \tag{2}$$

where h equals Planck's constant, c equals velocity of light; and as

$$n = \frac{\rho}{\mu H (1+f)} , \tag{3}$$

---

[1] II, No. 70, 592, 1931.

[2] *Philosophical Magazine*, 7, 63, 1929.

세상에서 가장 쉬운 과학 수업 별의 물리학

μ equals the molecular weight, 2.5, for a fully ionized material, H equals the mass of hydrogen atom, and f equals the ratio of number of ions to number of electrons, a factor usually negligible. Or, putting in the numerical values,

$$P = K\rho^{4/3}, \tag{4}$$

where K equals 3.619×10^{14}. We can now immediately apply the theory of polytropic gas spheres for the equation of state given by (4), where for the exponent γ we have

$$\gamma = \frac{4}{3} \text{ or } 1 + \frac{1}{n} = \frac{4}{3} \text{ or } n = 3 .$$

We have therefore the relation[1]

$$\left(\frac{GM}{M'}\right)^2 = \frac{(4K)^3}{4\pi G} ,$$

or

$$M = 1.822 \times 10^{33}$$
$$= .91\odot \text{ (nearly)} \tag{5}$$

As we have derived this mass for the star under ideal conditions of extreme degeneracy, we can regard $1.8_{22} \times 10^{33}$ as the maximum mass of an ideal white dwarf. This can be compared with the earlier estimate of Stoner[2]

$$M_{max} = 2.2 \times 10^{33}, \tag{6}$$

based again on uniform density distribution. The "agreement" between the accurate working out, based on the theory of the polytropes, and the cruder form of the theory is rather surprising in view of the fact that in the corresponding non-relativistic case the deviations were rather serious.

TRINITY COLLEGE
CAMBRIDGE
November 12, 1930

[1]A. S. Eddington, *Internal Constitution of Stars*, p. 83, eq. (57.3.)
[2]*Philosophical Magazine*, **9**, 944, 1930.

논문 웹페이지

위대한 논문과의 만남을 마무리하며

이 책은 별의 죽음 이론으로 별의 물리학을 완성한 찬드라세카르의 논문에 초점을 맞추었습니다. 더불어 이 논문이 나올 수 있게 별의 물리학을 연구한 레인, 엠덴, 에딩턴, 베테의 논문도 조금씩 다루어 보았습니다.

별의 죽음에 관한 논문을 소개하기에 앞서 우리는 고대 과학자들의 별에 대한 생각부터 망원경이 발명된 후의 별 관측 역사까지 살펴보았습니다. 또한 별을 물리학적으로 처음 논할 수 있게 해준 레인-엠덴의 연구를 수식으로 자세하게 설명했습니다. 비록 이 부분과 마지막 장의 찬드라세카르의 논문 해설에서 수식이 많이 나오지만, 그 외 부분에서는 수식을 되도록 멀리하고 별에 도전한 과학자들의 역사를 이야기했습니다.

사실 별의 물리학이나 천체물리학은 일부 대학을 제외하고는 대부분의 물리학과 학부 커리큘럼에 없는 과목입니다. 그렇지만 많은 대학생이 별의 신비와 별의 물리학에 궁금증을 가질 것이라 생각하여 이 시리즈에서 별의 죽음의 한 형태인 백색 왜성에 대한 찬드라세카르의 논문을 다루었습니다.

이 책의 출판 기획상 수식을 피할 수 없을 때는 고등학교 수학 정도를 아는 사람이라면 이해하도록 처음 쓴 원고를 고치고 또 고치는 작업을 반복했습니다. 그렇게 하여 수식을 줄여보려고 했습니다. 하지만 물리를 좋아하는 사람들이 쉽게 따라갈 수 있도록 친절하게 설

세상에서 가장 쉬운 과학 수업 별의 물리학

명했습니다.

원고를 쓰기 위해 19세기와 20세기 초의 여러 논문을 뒤적거렸습니다. 지금과는 완연히 다른 용어와 기호 때문에 많이 힘들었습니다. 특히 번역이 안 되어 있는 자료들이 많았지만 프랑스 논문에 대해서는 불문과를 졸업한 아내의 도움으로 조금은 이해할 수 있었습니다.

집필을 끝내자마자 다시 반도체를 이용해 트랜지스터를 발명한 바딘의 오리지널 논문을 공부하며, 시리즈를 계속 이어나갈 생각을 하니 즐거움에 벅차오릅니다. 제가 느끼는 이 기쁨을 독자들이 공유할 수 있기를 바라며 이제 힘들었지만 재미있었던 별에 관한 논문들과의 씨름을 여기서 멈추려고 합니다.

끝으로 용기를 내서 이 시리즈의 출간을 결정해준 성림원북스의 이성림 사장과 직원들에게 감사를 드립니다. 시리즈 초안이 나왔을 때, 수식이 많아 출판사들이 꺼릴 것 같다는 생각이 들었습니다. 몇 군데에 출판을 의뢰한 후 거절당하면 블로그에 올릴 생각으로 글을 써 내려갔습니다. 놀랍게도 첫 번째로 이 원고의 이야기를 나눈 성림원북스에서 출간을 결정해 주어서 책이 나올 수 있게 되었습니다. 원고를 쓰는 데 필요한 프랑스 논문의 번역을 도와준 아내에게도 고마움을 전합니다. 그리고 이 책을 쓸 수 있도록 멋진 논문을 만든 고 찬드라세카르 박사님에게도 감사를 드립니다.

진주에서 정완상 교수

이 책을 위해 참고한 논문들

1장

[1] N. Pogson, "Magnitudes of Thirty-six of the Minor Planets for the first day of each month of the year 1857", Monthly Notices of the Royal Astronomical Society. 17; 12-15, 1856.

2장

[1] I. Newton, Philosophiæ Naturalis Principia Mathematica, 1687.

[2] RJE. Clausius, "On a Mechanical Theorem Applicable to Heat", Philosophical Magazine. Series 4. 40 (265); 122-127, 1870.

[3] J. Bradley, "A Letter from the Reverend Mr. James Bradley Savilian Professor of Astronomy at Oxford, and F.R.S. to Dr. Edmond Halley Astronom. Reg. &c. Giving an Account of a New Discovered Motion of the Fix'd Stars", Philosophical Transactions of the Royal Society of London. 35; 637-661, 1728.

3장

[1] J. H. Lane, "On the theoretical temperature of the Sun, under

세상에서 가장 쉬운 과학 수업 별의 물리학

the hypothesis of a gaseous mass maintaining its volume by its internal heat, and depending on the laws of gases as known to terrestrial experiment", American Journal of Science. 2; 57-74, 1870.

[2] J. R. Emden, Gaskugeln: Anwendungen der mechanischen Wärmetheorie auf kosmologische und meteorologische probleme (Gas balls: Applications of the mechanical heat theory to cosmological and meteorological problems), 1907.

[3] A. S. Eddington, "On the radiative equilibrium of the stars", Monthly Notices of the Royal Astronomical Society. 77; 16-35, 1916.

[4] A. H. Compton, "A Quantum Theory of the Scattering of X−Rays by Light Elements", Physical Review. 21 (5); 483-502, May 1923.

4장

[1] E. Hertzsprung, "On the Use of Photographic Effective Wavelengths for the Determination of Color Equivalents", Publications of the Astrophysical Observatory in Potsdam. 1; 22, 1911.

[2] E. Hertzsprung, "Über die räumliche Verteilung der Veränderlichen vom δ Cephei−Typus", Astronomische

Nachrichten. 196; 201-208, 1913.

[3] H. N. Russell, "Relations Between the Spectra and Other Characteristics of the Stars", Popular Astronomy. 22; 275-294, 1914.

[4] J. H. Jeans, "The Stability of a Spherical Nebula", Philosophical Transactions of the Royal Society A. 199; 1-53, 1902.

[5] A. S. Eddington, The Internal Constitution of the Stars, The Scientific Monthly. 11; 297-303, 1920.

[6] R. Atkinson and F. G. Houtermans, "Zur Frage der Aufbaumöglichkeit der Elemente in Sternen", Zeitschrift für Physik. 54; 656-665, 1929.

[7] J. Chadwick, "Possible Existence of a Neutron", Nature. 129; 312, 1932.

[8] H. A. Bethe, "Energy Production in Stars", Physical Review. 55; 434-456, 1939.

5장

[1] S. Chandrasekhar, "The Density of White Dwarf Stars", Philosophical Magazine. 11 (70); 592-596, 1931.

[2] S. Chandrasekhar, "The maximum mass of ideal white dwarfs", The Astrophysical Journal. 74 (1); 81–82, 1931.

수식에 사용하는 그리스 문자

대문자	소문자	읽기	대문자	소문자	읽기
A	α	알파(alpha)	N	ν	뉴(nu)
B	β	베타(beta)	Ξ	ξ	크시(xi)
Γ	γ	감마(gamma)	O	o	오미크론(omicron)
Δ	δ	델타(delta)	Π	π	파이(pi)
E	ε	엡실론(epsilon)	P	ρ	로(rho)
Z	ζ	제타(zeta)	Σ	σ	시그마(sigma)
H	η	에타(eta)	T	τ	타우(tau)
Θ	θ	세타(theta)	Y	υ	입실론(upsilon)
I	ι	요타(iota)	Φ	φ	피(phi)
K	χ	카파(kappa)	X	χ	키(chi)
Λ	λ	람다(lambda)	Ψ	ψ	프시(psi)
M	μ	뮤(mu)	Ω	ω	오메가(omega)

노벨 물리학상 수상자들을 소개합니다

이 책에 언급된 노벨상 수상자는 이름 앞에 ★로 표시하였습니다.

연도	수상자	수상 이유
1901	빌헬름 콘라트 뢴트겐	그의 이름을 딴 놀라운 광선의 발견으로 그가 제공한 특별한 공헌을 인정하여
1902	헨드릭 안톤 로런츠 피터르 제이만	복사 현상에 대한 자기의 영향에 대한 연구를 통해 그들이 제공한 탁월한 공헌을 인정하여
1903	앙투안 앙리 베크렐	자발 방사능 발견으로 그가 제공한 탁월한 공로를 인정하여
1903	피에르 퀴리 마리 퀴리	앙리 베크렐 교수가 발견한 방사선 현상에 대한 공동 연구를 통해 그들이 제공한 탁월한 공헌을 인정하여
1904	★존 윌리엄 스트럿 레일리	가장 중요한 기체의 밀도에 대한 조사와 이러한 연구와 관련하여 아르곤을 발견한 공로
1905	필리프 레나르트	음극선에 대한 연구
1906	조지프 존 톰슨	기체에 의한 전기 전도에 대한 이론적이고 실험적인 연구의 큰 장점을 인정하여
1907	앨버트 에이브러햄 마이컬슨	광학 정밀 기기와 그 도움으로 수행된 분광 및 도량형 조사
1908	가브리엘 리프만	간섭 현상을 기반으로 사진적으로 색상을 재현하는 방법
1909	굴리엘모 마르코니 카를 페르디난트 브라운	무선 전신 발전에 기여한 공로를 인정받아
1910	요하네스 디데릭 판데르발스	기체와 액체의 상태 방정식에 관한 연구
1911	빌헬름 빈	열복사 법칙에 관한 발견
1912	닐스 구스타프 달렌	등대와 부표를 밝히기 위해 가스 어큐뮬레이터와 함께 사용하기 위한 자동 조절기 발명

세상에서 가장 쉬운 과학 수업 별의 물리학

1913	헤이커 카메를링 오너스	특히 액체 헬륨 생산으로 이어진 저온에서의 물질 특성에 대한 연구
1914	막스 폰 라우에	결정에 의한 X선 회절 발견
1915	윌리엄 헨리 브래그	X선을 이용한 결정구조 분석에 기여한 공로
	윌리엄 로런스 브래그	
1916	수상자 없음	
1917	찰스 글러버 바클라	원소의 특징적인 뢴트겐 복사 발견
1918	막스 플랑크	에너지 양자 발견으로 물리학 발전에 기여한 공로 인정
1919	요하네스 슈타르크	커낼선의 도플러 효과와 전기장에서 분광선의 분할 발견
1920	샤를 에두아르 기욤	니켈강 합금의 이상 현상을 발견하여 물리학의 정밀 측정에 기여한 공로를 인정하여
1921	★알베르트 아인슈타인	이론 물리학에 대한 공로, 특히 광전효과 법칙 발견
1922	닐스 보어	원자 구조와 원자에서 방출되는 방사선 연구에 기여
1923	로버트 앤드루스 밀리컨	전기의 기본 전하와 광전효과에 관한 연구
1924	칼 만네 예오리 시그반	X선 분광학 분야에서의 발견과 연구
1925	제임스 프랑크	전자가 원자에 미치는 영향을 지배하는 법칙 발견
	구스타프 헤르츠	
1926	장 바티스트 페랭	물질의 불연속 구조에 관한 연구, 특히 침전 평형 발견
1927	★아서 콤프턴	그의 이름을 딴 효과 발견
	찰스 톰슨 리스 윌슨	수증기 응축을 통해 전하를 띤 입자의 경로를 볼 수 있게 만든 방법
1928	오언 윌런스 리처드슨	열전자 현상에 관한 연구, 특히 그의 이름을 딴 법칙 발견
1929	루이 드브로이	전자의 파동성 발견
1930	★찬드라세카라 벵카타 라만	빛의 산란에 관한 연구와 그의 이름을 딴 효과 발견
1931	수상자 없음	

1932	베르너 하이젠베르크	수소의 동소체 형태 발견으로 이어진 양자역학의 창시
1933	★에르빈 슈뢰딩거	원자 이론의 새로운 생산적 형태 발견
	폴 디랙	
1934	수상자 없음	
1935	★제임스 채드윅	중성자 발견
1936	빅토르 프란츠 헤스	우주 방사선 발견
	★칼 데이비드 앤더슨	양전자 발견
1937	클린턴 조지프 데이비슨	결정에 의한 전자의 회절에 대한 실험적 발견
	조지 패짓 톰슨	
1938	★엔리코 페르미	중성자 조사에 의해 생성된 새로운 방사성 원소의 존재에 대한 시연 및 이와 관련된 느린중성자에 의한 핵반응 발견
1939	어니스트 로런스	사이클로트론의 발명과 개발, 특히 인공 방사성 원소와 관련하여 얻은 결과
1940	수상자 없음	
1941		
1942		
1943	오토 슈테른	분자선 방법 개발 및 양성자의 자기 모멘트 발견에 기여
1944	이지도어 아이작 라비	원자핵의 자기적 특성을 기록하기 위한 공명 방법
1945	★볼프강 파울리	파울리 원리라고도 불리는 배제 원리의 발견
1946	퍼시 윌리엄스 브리지먼	초고압을 발생시키는 장치의 발명과 고압 물리학 분야에서 그가 이룬 발견에 대해
1947	에드워드 빅터 애플턴	대기권 상층부의 물리학 연구, 특히 이른바 애플턴층의 발견
1948	패트릭 메이너드 스튜어트 블래킷	윌슨 구름상자 방법의 개발과 핵물리학 및 우주 방사선 분야에서의 발견
1949	유카와 히데키	핵력에 관한 이론적 연구를 바탕으로 중간자 존재 예측

세상에서 가장 쉬운 과학 수업 별의 물리학

1950	세실 프랭크 파월	핵 과정을 연구하는 사진 방법의 개발과 이 방법으로 만들어진 중간자에 관한 발견
1951	존 더글러스 콕크로프트	인위적으로 가속된 원자 입자에 의한 원자핵 변환에 대한 선구자적 연구
	어니스트 토머스 신턴 월턴	
1952	펠릭스 블로흐	핵자기 정밀 측정을 위한 새로운 방법 개발 및 이와 관련된 발견
	에드워드 밀스 퍼셀	
1953	프리츠 제르니커	위상차 방법 시연, 특히 위상차 현미경 발명
1954	막스 보른	양자역학의 기초 연구, 특히 파동함수의 통계적 해석
	발터 보테	우연의 일치 방법과 그 방법으로 이루어진 그의 발견
1955	윌리스 유진 램	수소 스펙트럼의 미세 구조에 관한 발견
	폴리카프 쿠시	전자의 자기 모멘트를 정밀하게 측정한 공로
1956	윌리엄 브래드퍼드 쇼클리	반도체 연구 및 트랜지스터 효과 발견
	존 바딘	
	월터 하우저 브래튼	
1957	★양전닝	소립자에 관한 중요한 발견으로 이어진 소위 패리티 법칙에 대한 철저한 조사
	★리정다오	
1958	파벨 알렉세예비치 체렌코프	체렌코프 효과의 발견과 해석
	일리야 프란크	
	이고리 탐	
1959	에밀리오 지노 세그레	반양성자 발견
	오언 체임벌린	
1960	도널드 아서 글레이저	거품 상자의 발명
1961	로버트 호프스태터	원자핵의 전자 산란에 대한 선구적인 연구와 핵자 구조에 관한 발견
	루돌프 뫼스바워	감마선의 공명 흡수에 관한 연구와 그의 이름을 딴 효과에 대한 발견

1962	레프 다비도비치 란다우	응집 물질, 특히 액체 헬륨에 대한 선구적인 이론
1963	유진 폴 위그너	원자핵 및 소립자 이론에 대한 공헌, 특히 기본 대칭 원리의 발견 및 적용을 통한 공로
	마리아 괴페르트 메이어	핵 껍질 구조에 관한 발견
	한스 옌젠	
1964	니콜라이 바소프	메이저-레이저 원리에 기반한 발진기 및 증폭기의 구성으로 이어진 양자 전자 분야의 기초 작업
	알렉산드르 프로호로프	
	찰스 하드 타운스	
1965	도모나가 신이치로	소립자의 물리학에 심층적인 결과를 가져온 양자전기역학의 근본적인 연구
	줄리언 슈윙거	
	리처드 필립스 파인먼	
1966	알프레드 카스틀레르	원자에서 헤르츠 공명을 연구하기 위한 광학적 방법의 발견 및 개발
1967	★한스 알브레히트 베테	핵반응 이론, 특히 별의 에너지 생산에 관한 발견에 기여
1968	루이스 월터 앨버레즈	소립자 물리학에 대한 결정적인 공헌, 특히 수소 기포 챔버 사용 기술 개발과 데이터 분석을 통해 가능해진 다수의 공명 상태 발견
1969	머리 겔만	기본 입자의 분류와 그 상호 작용에 관한 공헌 및 발견
1970	한네스 올로프 예스타 알벤	플라스마 물리학의 다양한 부분에서 유익한 응용을 통해 자기유체역학의 기초 연구 및 발견
	루이 외젠 펠릭스 네엘	고체 물리학에서 중요한 응용을 이끈 반강자성 및 강자성에 관한 기초 연구 및 발견
1971	데니스 가보르	홀로그램 방법의 발명 및 개발
1972	존 바딘	일반적으로 BCS 이론이라고 하는 초전도 이론을 공동으로 개발한 공로
	리언 닐 쿠퍼	
	존 로버트 슈리퍼	

	에사키 레오나	반도체와 초전도체의 터널링 현상에 관한 실험적 발견
1973	이바르 예베르	
	브라이언 데이비드 조지프슨	터널 장벽을 통과하는 초전류 특성, 특히 일반적으로 조지프슨 효과로 알려진 현상에 대한 이론적 예측
1974	★마틴 라일	전파 천체물리학의 선구적인 연구: 라일은 특히 개구 합성 기술의 관찰과 발명, 그리고 휴이시는 펄서 발견에 결정적인 역할을 함
	★앤터니 휴이시	
	오게 닐스 보어	원자핵에서 집단 운동과 입자 운동 사이의 연관성 발견과 이 연관성에 기초한 원자핵 구조 이론 개발
1975	벤 로위 모텔손	
	제임스 레인워터	
1976	버턴 릭터	새로운 종류의 무거운 기본 입자 발견에 대한 선구적인 작업
	새뮤얼 차오 충 팅	
	필립 워런 앤더슨	자기 및 무질서 시스템의 전자 구조에 대한 근본적인 이론적 조사
1977	네빌 프랜시스 모트	
	존 해즈브룩 밴블렉	
	표트르 레오니도비치 카피차	저온 물리학 분야의 기본 발명 및 발견
1978	아노 앨런 펜지어스	우주 마이크로파 배경 복사의 발견
	로버트 우드로 윌슨	
	셸던 리 글래쇼	특히 약한 중성 전류의 예측을 포함하여 기본 입자 사이의 통일된 약한 전자기 상호 작용 이론에 대한 공헌
1979	압두스 살람	
	스티븐 와인버그	
1980	제임스 왓슨 크로닌	중성 K 중간자의 붕괴에서 기본 대칭 원리 위반 발견
	밸 로그즈던 피치	
	니콜라스 블룸베르헌	레이저 분광기 개발에 기여
1981	아서 레너드 숄로	
	카이 만네 뵈리에 시그반	고해상도 전자 분광기 개발에 기여

1982	케네스 게디스 윌슨	상전이와 관련된 임계 현상에 대한 이론
1983	★수브라마니안 찬드라세카르	별의 구조와 진화에 중요한 물리적 과정에 대한 이론적 연구
	윌리엄 앨프리드 파울러	우주의 화학 원소 형성에 중요한 핵반응에 대한 이론 및 실험적 연구
1984	카를로 루비아	약한 상호 작용의 커뮤니케이터인 필드 입자 W와 Z의 발견으로 이어진 대규모 프로젝트에 결정적인 기여
	시몬 판데르 메이르	
1985	클라우스 폰 클리칭	양자화된 홀 효과의 발견
1986	에른스트 루스카	전자 광학의 기초 작업과 최초의 전자 현미경 설계
	게르트 비니히	스캐닝 터널링 현미경 설계
	하인리히 로러	
1987	요하네스 게오르크 베드노르츠	세라믹 재료의 초전도성 발견에서 중요한 돌파구
	카를 알렉산더 뮐러	
1988	리언 레더먼	뉴트리노 빔 방법과 뮤온 중성미자 발견을 통한 경입자의 이중 구조 증명
	멜빈 슈워츠	
	잭 스타인버거	
1989	노먼 포스터 램지	분리된 진동 필드 방법의 발명과 수소 메이저 및 기타 원자시계에서의 사용
	한스 게오르크 데멜트	이온 트랩 기술 개발
	볼프강 파울	
1990	제롬 프리드먼	입자 물리학에서 쿼크 모델 개발에 매우 중요한 역할을 한 양성자 및 구속된 중성자에 대한 전자의 심층 비탄성 산란에 관한 선구적인 연구
	헨리 웨이 켄들	
	리처드 테일러	
1991	피에르질 드젠	간단한 시스템에서 질서 현상을 연구하기 위해 개발된 방법을 보다 복잡한 형태의 물질, 특히 액정과 고분자로 일반화할 수 있음을 발견

세상에서 가장 쉬운 과학 수업 별의 물리학

1992	조르주 샤르파크	입자 탐지기, 특히 다중 와이어 비례 챔버의 발명 및 개발
1993	러셀 헐스	새로운 유형의 펄서 발견, 중력 연구의 새로운 가능성을 연 발견
	조지프 테일러	
1994	버트럼 브록하우스	중성자 분광기 개발
	클리퍼드 셜	중성자 회절 기술 개발
1995	마틴 펄	타우 렙톤의 발견
	★프레더릭 라이너스	중성미자 검출
1996	데이비드 리	헬륨-3의 초유동성 발견
	더글러스 오셔로프	
	로버트 리처드슨	
1997	스티븐 추	레이저 광으로 원자를 냉각하고 가두는 방법 개발
	클로드 코엔타누지	
	윌리엄 필립스	
1998	로버트 로플린	부분적으로 전하를 띤 새로운 형태의 양자 유체 발견
	호르스트 슈퇴르머	
	대니얼 추이	
1999	헤라르뒤스 엇호프트	물리학에서 전기약력 상호작용의 양자 구조 규명
	마르티뉘스 펠트만	
2000	조레스 알표로프	정보 통신 기술에 대한 기초 작업(고속 및 광전자 공학에 사용되는 반도체 이종 구조 개발)
	허버트 크로머	
	잭 킬비	정보 통신 기술에 대한 기초 작업(집적 회로 발명에 기여)
2001	에릭 코넬	알칼리 원자의 희석 가스에서 보스-아인슈타인 응축 달성 및 응축 특성에 대한 초기 기초 연구
	칼 위먼	
	볼프강 케테를레	

2002	레이먼드 데이비스	천체물리학, 특히 우주 중성미자 검출에 대한 선구적인 공헌
	고시바 마사토시	
	리카르도 자코니	우주 X선 소스의 발견으로 이어진 천체물리학에 대한 선구적인 공헌
2003	알렉세이 아브리코소프	초전도체 및 초유체 이론에 대한 선구적인 공헌
	비탈리 긴즈부르크	
	앤서니 레깃	
2004	데이비드 그로스	강한 상호작용 이론에서 점근적 자유의 발견
	데이비드 폴리처	
	프랭크 윌첵	
2005	로이 글라우버	광학 일관성의 양자 이론에 기여
	존 홀	광 주파수 콤 기술을 포함한 레이저 기반 정밀 분광기 개발에 기여
	테오도어 헨슈	
2006	존 매더	우주 마이크로파 배경 복사의 흑체 형태와 이방성 발견
	조지 스무트	
2007	알베르 페르	자이언트 자기 저항의 발견
	페터 그륀베르크	
2008	난부 요이치로	아원자 물리학에서 자발적인 대칭 깨짐 메커니즘 발견
	고바야시 마코토	자연계에 적어도 세 종류의 쿼크가 존재함을 예측하는 깨진 대칭의 기원 발견
	마스카와 도시히데	
2009	찰스 가오	광 통신을 위한 섬유의 빛 전송에 관한 획기적인 업적
	윌러드 보일	영상 반도체 회로(CCD 센서)의 발명
	조지 엘우드 스미스	
2010	안드레 가임	2차원 물질 그래핀에 관한 획기적인 실험
	콘스탄틴 노보셀로프	

2011	솔 펄머터	원거리 초신성 관측을 통한 우주 가속 팽창 발견
	브라이언 슈밋	
	애덤 리스	
2012	세르주 아로슈	개별 양자 시스템의 측정 및 조작을 가능하게 하는 획기적인 실험 방법
	데이비드 와인랜드	
2013	프랑수아 앙글레르	아원자 입자의 질량 기원에 대한 이해에 기여하고 최근 CERN의 대형 하드론 충돌기에서 ATLAS 및 CMS 실험을 통해 예측된 기본 입자의 발견을 통해 확인된 메커니즘의 이론적 발견
	피터 힉스	
2014	아카사키 이사무	밝고 에너지 절약형 백색 광원을 가능하게 한 효율적인 청색 발광 다이오드의 발명
	아마노 히로시	
	나카무라 슈지	
2015	가지타 다카아키	중성미자가 질량을 가지고 있음을 보여주는 중성미자 진동 발견
	아서 맥도널드	
2016	데이비드 사울레스	위상학적 상전이와 물질의 위상학적 위상에 대한 이론적 발견
	덩컨 홀데인	
	마이클 코스털리츠	
2017	라이너 바이스	LIGO 탐지기와 중력파 관찰에 결정적인 기여
	킵 손	
	배리 배리시	
2018	아서 애슈킨	레이저 물리학 분야의 획기적인 발명(광학 핀셋과 생물학적 시스템에 대한 응용)
	제라르 무루	레이저 물리학 분야의 획기적인 발명(고강도 초단파 광 펄스 생성 방법)
	도나 스트리클런드	
2019	제임스 피블스	우주의 진화와 우주에서 지구의 위치에 대한 이해에 기여(물리 우주론의 이론적 발견)
	미셸 마요르	우주의 진화와 우주에서 지구의 위치에 대한 이해에 기여(태양형 항성 주위를 공전하는 외계 행성 발견)
	디디에 쿠엘로	

2020	로저 펜로즈	블랙홀 형성이 일반 상대성 이론의 확고한 예측이라는 발견
	★라인하르트 겐첼	우리 은하의 중심에 있는 초거대 밀도 물체 발견
	앤드리아 게즈	
2021	마나베 슈쿠로	복잡한 시스템에 대한 이해에 획기적인 기여(지구 기후의 물리적 모델링, 가변성을 정량화하고 지구 온난화를 안정적으로 예측)
	클라우스 하셀만	
	조르조 파리시	복잡한 시스템에 대한 이해에 획기적인 기여 (원자에서 행성 규모에 이르는 물리적 시스템의 무질서와 요동의 상호작용 발견)
2022	알랭 아스페	얽힌 광자를 사용한 실험, 벨 불평등 위반 규명 및 양자 정보 과학 개척
	존 클라우저	
	안톤 차일링거	
2023	피에르 아고스티니	물질의 전자 역학 연구를 위해 아토초(100경분의 1초) 빛 펄스를 생성하는 실험 방법 고안
	페렌츠 크러우스	
	안 륄리에	

세상에서 가장 쉬운 과학 수업 별의 물리학